The Ministry of Nature

Harold W. Clark, MA

Director
Pacific Union College Field Nature School

Author
Genesis and Science
Genes and Genesis

TEACH Services, Inc.
P U B L I S H I N G
www.TEACHServices.com • (800) 367-1844

World rights reserved. This book or any portion thereof may not be copied or reproduced in any form or manner whatever, except as provided by law, without the written permission of the publisher, except by a reviewer who may quote brief passages in a review.

The opinions expressed in this book are the author's personal views and interpretations, and do not necessarily reflect those of the publisher.

This book is provided with the understanding that the publisher is not engaged in giving spiritual, legal, medical, or other professional advice. If authoritative advice is needed, the reader should seek the counsel of a competent professional.

Copyright 2018 © TEACH Services, Inc.
Copyright © 1941 Harold W. Clark

ISBN-13: 978-1-4796-0804-1 (Paperback)
Library of Congress Control Number: 2018939498

Illustrations by Wm. E. Steinbach

Image Credits:
Cover, lauritta/Bigstock; p. 4a, granitepeaker/Bigstock; p. 4b, reisegraf.ch/Bigstock; p. 36a, lucky-photographer/Bigstock; p. 36b, _jure/Bigstock; p. 68a-top, Andrey Armyagov/Bigstock; p. 68a-bottom, suthisa/Bigstock; p. 68b-top, lbryan/Bigstock; p. 68b-bottom, Sergey Novikov/Bigstock; p. 100a, Erin Donalson/Bigstock; p. 100b, DmitryP/Bigstock

TEACH Services, Inc.
P U B L I S H I N G
www.TEACHServices.com • (800) 367-1844

DEDICATED

To Melva June, whose interest in nature from childhood has always been a source of comfort and encouragement to the author, her father, as he has seen in her the development of the principles of nature study for which he has been contending, this work is affectionately dedicated.

> *"Come forth into the light of things,
> Let Nature be your teacher."*
> —*Wordsworth.*

Preface

This volume has been prepared in answer to many urgent requests from teachers and other workers. While there is abundant factual material available for nature teaching, there is a dearth of illustrative material for bringing out the spiritual lessons which nature was originally designed to teach. This book is sent out with the hope that it may help to fill the need.

There is no attempt to make the subject matter exhaustive, for the obvious reason that nature is infinite in its scope, and no amount of study could ever exhaust its lessons. Since it is the expression of the thoughts of God, there will ever be an eternity of study before the nature student, and yet he will ever find new wonders to explore, new mysteries to solve. All that can be hoped is that the lessons suggested in the following pages may point the way to a fuller understanding of the truths revealed in the great book of nature.

In the preparation of this material the author has consulted many works of a technical nature, dealing with all phases of science and nature study; he has also drawn from the writings of several world-renowned authors, among whom are the names of Ellen G. White, Hugh Macmillan, Henry Drummond, Charles Spurgeon, and others. From miscellaneous sources valuable help has been obtained.

The author wishes to thank those of his associates in nature work during the past years for their encouragement and sympathetic fellowship that has made this book possible. Many of them will never realize that they have had any part in it. George Jeys, Kenneth Farnsworth, Joseph Hatton, Cornell McReynolds, Donald Hemphill, Ernest Booth, Mr. and Mrs. Herschel Wheeler, Mrs. S. J. Whitney, Mr. and Mrs. Charles Weniger, and many others, especially members of the Pacific Union College Field Nature School year by year,—have all contributed their part. Thanks are due to my patient wife, who has followed me into many a wild region in search of nature secrets, and who has given of her time to type the manuscript from my illegible scrawls.

HAROLD W. CLARK.

Angwin, California
March 1, 1941.

How to Use the Lessons

All the material in this book has been arranged in alphabetical, or encyclopedic, order. A system of cross-references makes it easy to study related topics. A special feature is the manner in which related items are correlated. For example, under the heading ANIMALS, are collected, in alphabetical order, all the references to individual animals, such as Bear, Lion, etc. These are also listed in their regular place, with a cross-reference to the general topic where they are given. The same plan is followed with respect to BIRDS, FLOWERS, INSECTS, TREES, and other major topics.

The discussions are not lengthy. On the contrary, they are made very brief in order to include the largest number possible in the space available. They are intended more as suggestions, upon which the reader may expand, and from which he may draw ideas for fuller study and discussion.

These lessons may be used by parents, teachers, Sabbath-school and young people's workers, and by ministers. Teachers will find them helpful in nature classes and for opening exercises. They may be used in Sabbath school to bring out practical lessons. A very good way to use them is to bring in some nature object and then draw the lesson from it. Young people's missionary meetings, or group meetings of any kind, where some inspirational lesson is desired, may draw from this source. Parents will find the suggestions helpful in teaching their children to love and appreciate nature. The minister in the pulpit may take one of the suggested lessons and expand upon it or draw an object lesson from it. Such lessons are often more effective than the stories that are so commonly used in the pulpit.

The Ministry of Nature

In Eden there was nothing to obscure nature's lessons. The existence of God, His wonder-working power, and His boundless love, were clearly seen in the objects of nature. All created things were an expression of the thoughts of God. The whole natural world was full of lessons of divine truth. It was His purpose that man should hear and understand the communications of God by contact with created things.

God's plan has never changed, even though sin has obscured many of the lessons. Nature still speaks of God, and upon every leaf and flower and tree is written the message of His love. The beauties of earth and sea and sky tell of His matchless glory. And God would still have His children learn to love and appreciate His handiwork. Through a diligent study of nature they are to obtain a knowledge of His character.

In order to be more than a mere collection of factual material, nature study must be vivified by the Spirit of God. Since it represents the thought of God expressed in outward manifestation, its proper study ought to bring forth the spirit that is within. We ought to approach the study of nature in a reverential manner, asking God to reveal to our hearts the spiritual truth that He would have us learn from our study of the works of creation.

There are four ways in which nature study contributes to the spiritual life. These are more or less interwoven, yet they may be considered separately.

1. The Bible is full of the imagery of nature. Frequent references are made to nature objects and phenomena,—to trees, birds, animals, flowers, the sun, moon, stars, rain, hail, snow, dew, and many others. Deep spiritual truths are illustrated, and nature parables are used to teach principles that could not be understood if dealt with abstractly. The student of nature will find in the Bible many suggestions regarding the spiritual value of

nature study, and will be able to adapt many of its nature parables to his own life and experience.

2. Nature itself is full of object lessons. The open eye and the listening ear will be quick to detect many of these. The most common things may be made to teach beautiful truths, if we are wise to discern their meaning. The lessons are never trite, but are always new, ready to be adapted to the particular situation in which we find ourselves. A nature lesson may have one application today and another tomorrow. If we learn how to read nature's truths, we shall be able to draw continually from this source. There is no limit to the number and variety of nature lessons that may be taught. As nature is the expression of an infinite God, its lessons are infinite in their scope.

3. The intricate adaptations in nature teach more of God than any theoretical discussions can do. "God is Love" is written not only in the Word, but on every leaf of the forest and every spear of grass. The way in which this earth has been fitted as a habitation for living creatures reveals God's love for His children. The beauties of earth, sea, and sky show His character. The marvelous manner in which all living things are able to maintain their existence teaches us of His care for them.

We must learn to recognize these truths and to appreciate their value in a spiritual way. It is a more effective nature lesson to find, for example, how God adapts the structure of the leaf to the changes in the atmosphere, than to try to make a vague parable from the leaf. In other words, the very existence of the complex relations in nature, by which life is maintained, is as great a truth as can be found. True scientific study recognizes God as the creator and upholder of all things, and sees in all nature the outworking of His power.

4. Nature speaks directly to the heart that is open to the influence of God's spirit. The hardened sinner may see no spiritual truth in nature, because his heart is closed to truth. But anyone who is willing to receive the lessons of nature will find in nature,—the beauties of sunrise and sunset, the changing aspects of the heavens, the loveliness of the flowers, the majesty of the trees,—an unending source of uplift and spiritual enlightenment.

The mind must cultivate a love for the beautiful, and an appreciation for that which is grand and ennobling. While words may be totally inadequate to express the feelings that come from contact with natural loveliness, its influence is none the less real.

Often there is more value in a silent observation of some majestic natural phenomenon than in all the voluble lessons that might be developed about it. Teachers need to learn when to speak, when to make only suggestions, and when to keep silent and let God speak through His own appointed agencies. If the children are taught to recognize the voice of God in nature, He will speak to them many times when our dull senses would fail to detect His presence.

In a way not generally recognized, nature offers a solution to the problem of recreation. Instead of devising artificial methods of entertainment,—parties, indoor games, etc.,—teachers and leaders would do well to organize outdoor activities. Hikes, camping out, trips to points of natural interest, collecting flowers, minerals, etc., will afford an outlet for the restless energy of youth, and will not leave the undesirable after-effects that come from artificial amusements. When once an interest has been created in outdoor activities, the spiritual influences of nature will become powerful factors in character development.

Christ's method of teaching should receive more attention in educational work.

"So wide was Christ's view of truth, so extended His teaching, that every phase of nature was employed in illustrating truth. The scenes upon which the eye daily rests were all connected with some spiritual truth, so that nature is clothed with the parables of the Master.

"Christ sought to remove that which obscured the truth. The veil that sin has cast over the face of nature, He came to draw aside, bringing to view the spiritual glory that all things were created to reflect.

"Jesus desired to awaken inquiry. He sought to arouse the careless, and impress truth upon the heart.

"By connecting His teaching with the scenes of life, experience, and nature, He secured their attention and impressed their hearts. Afterward, as they looked upon the objects that illustrated His lessons, they recalled the words of the divine Teacher. . . . Mysteries grew clear, and that which had been hard to grasp became evident.

"So we should teach. Let the children learn to see in nature an expression of the love and wisdom of God; let the thought of Him be linked with bird and flower and tree; let all things seen become to them the interpretation of the unseen, and all the events of life be a means of divine teaching.

"Let everything which their eyes see or their hands handle be made a lesson in character building. Thus the mental powers will be strengthened, and the character developed, the whole life ennobled.

"Teach the children to see Christ in nature. Take them out into the open air, under the noble trees, into the garden; and in all the wonderful works of creation teach them to see an expression of His love. . . . Do not weary them with long prayers and tedious exhortations, but through nature's object-lessons teach them obedience to the law of God."—*Ellen G. White, Miscellaneous Selections.*

What lessons are of most importance? How may we get these lessons across to our children and to others? A few suggestions may be of help.

1. *Teach the universal dominion of law.*— Nature's laws are God's laws. Everything from the tiny mote that floats in the sunbeam to the mightiest creature of the sea is under law, and must obey the laws of its being in order to survive. Happiness is to be had only by obedience to the laws of our existence,— physical, moral, and spiritual.

2. *Teach the unity of man with nature and with God.*— Every living thing receives the sunlight, air, and food provided by God, and in turn contributes to the support of other forms of life. The great scientific principle of adaptation teaches us how living things are fitted to their environment, and how they maintain an existence amid the changes that go on around them.

3. *Teach the wonder and marvel of the universe,*— and especially of living plants and animals. To understand the life pro-

cesses that go on in nature requires the deepest thought and excites the highest admiration. By innumerable examples the Creator has shown us His power. By studying His care for His creatures we are filled with love for Him.

4. *Teach appreciation for the beauties of nature.*— Man has been endowed with powers of appreciation of music, color, and artistic form. Art itself, in music, painting, sculpture, or any other mode of expression, is a reflection of the beauty of nature. Our lives need to be enriched by a wealth of beauty with which we are altogether too little acquainted.

5. *Teach the law of service.*— No living thing exists for itself. Each must pass on its blessings to others, in order to be successful in its own life.

There are endless variations that might be built upon these five points. But in brief we might summarize the objectives in nature teaching as follows:

1. To bring about a fuller understanding of the character of God, by observing the manner in which He deals with the creatures of His hand.

2. To show the wisdom, love, and power of God, and point the way to communion with Him.

3. To develop a love and appreciation for the beautiful and true, that cheap worldly attractions may find no place in the life.

4. To make eternal truth the heritage of every person who can be reached by it, that he may pass it on to others.

5. To stimulate the imagination, and cultivate the powers of body, soul, and spirit.

6. To develop a character in harmony with the principles of truth as revealed in the Word and works of God.

Nature Lessons

ADDER,—See Reptiles.

AIR,—See also Clouds; Dew; Fire; Oxidation; Snow; Storm; Whirlwind; Wind.

Heated air rises. Thus a fresh supply is brought in to replace it. This is nature's "air-conditioning."

The movement of air produces a breeze. The result is a cooling effect.

Heated air from interior valleys rises up the mountain slopes. If the valley is near a lake or sea, cooling breezes come in to replace the hot air.

The heated air moving up the mountains, warms them, melts the snows in spring, and causes showers in summer.

If hot air settled, it would keep the valleys under stifling heat, and the mountains would never become warm.

Air has weight. Because of this, swiftly moving objects such as birds and insects, find sufficient resistance to enable them to fly.

Even tho the air has weight, it is not so heavy but that slow-moving creatures, such as man and animals, are not impeded in their movements.

Air transmits sound. Without it, hearing would be impossible. Communication could take place only by signs visible to the eye. Without the pressure of sound, all music and the voice would be impossible.

Air diffuses the rays of light, so that we obtain the soft light of the sky and the even light from the landscape. The light is scattered by the particles of water vapor and dust. Without these, the heavens would appear black, and objects in the shade would be absolutely invisible. The sun would shine with an unendurable glare upon everything. Not only light, but heat, would be terrific in the sun's rays, whereas the shadows would be icy cold. Thus the nature of the air provides wonderfully for life upon the earth.

If there were no air, the earth would be plunged into instant darkness at sunset, and would emerge from darkness into full light at sunrise. Wild creatures would have no warning of oncoming night, and no chance to find a place of repose for the night.

Air is the blanket to keep the earth warm. The radiant energy of the sun is of such short wave length that it passes through the air without heating it, or being changed by it. When these waves reach the earth, they are absorbed, and many of them are reflected into the air as heat waves.

Since the heat waves are shorter, they heat the air, and do not pass through it as readily as does light. Thus the earth is warmed as by a transparent blanket.

The water vapor in the air is a thermal regulator. A humid atmosphere is not quickly affected by heat and cold, for the latent heat of water is very high, and it requires a large amount of heat to change its temperature. By means of its blanket of water vapor the earth is protected from sudden changes in temperature. As an illustration we may notice the difference between the extremes in a desert and along the seashore. Sometimes in the Sahara it will be over 100° at midday and below freezing at night. At Eureka, California, the temperature hardly varies more than 4° summer and winter.

Air is heavier than water vapor. The molecular weight of water vapor is 18; that of air is about 29. This enables the water vapor to rise into the air. If it were not for this, the phenomena of clouds, rain, dew, and the like, would be impossible. Warm air holds more moisture than cold air. When warm air from the surface of large bodies of water or from the valleys passes over cool mountain slopes, it is driven into the upper atmosphere, and its load of water vapor is condensed into clouds or rain. The earth is watered by the principle of condensation.

If the air were much thinner than it is, life would be difficult, if not impossible, at high altitudes. If it were three or four times as dense, its force when moving would be so great that nothing could stand before it. The proper weight has been given to the air to make it the most suitable medium in which life may exist.

The air is made up of a mixture of oxygen and nitrogen in a proportion of roughly one to four. If the oxygen were pure, or even in a high proportion, combustion would occur at a low temperature. Even iron will burn when heated red-hot and plunged in an atmosphere of pure oxygen.

Nitrogen and oxygen are in a mixture, not as a compound. Compounds of nitrogen and oxygen are very poisonous.

The air that a man breathes either purifies or poisons his body. So the society that he keeps brings blessing or harmful influences into his life.

ALGAE,—See Plants.

ANATOMY,—See also Animals.

Studies in anatomy reveal many marvelous devices by which the bodies of man and animals are fitted for their life. The subject is too extensive for detailed treatment here. Only a few cases can be noted.

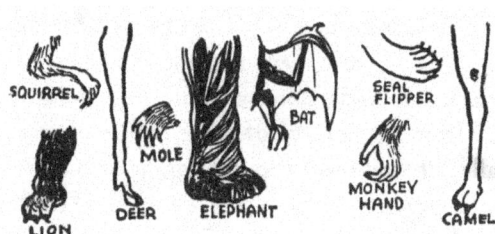

In animal skeletons are seen remarkable adaptations for the particular kinds of locomotion, of which the following are examples:

1. Deer and other similar animals have a thin body, with the legs set close to the sides. The lower leg and foot bones are long and slender, making great speed possible.

2. The elephant has short, heavy legs located underneath the body, in order to support its tremendous weight.

"Nature testifies of God." (*Education*, p. 99)

"...nature presents an unfailing source of instruction and delight." (*Education*, p. 100)

3. Flying animals, like the bat, have extremely long, slender front limbs, to which the wings are attached.

4. Burrowing animals, such as the mole or gopher, have short, powerful limbs.

5. Climbing animals, like the squirrel, have very flexible limbs, with long feet, and toes, fitted with long claws.

6. In swimming animals the limbs are modified to form fins or flippers.

In four-footed animals the elbows are bent backward and the knees forward. This makes a pull on the front leg and a push on the hind leg.

The front limbs of mammals are modified into various structures, such as wings in the bat, flippers in the seal, hands in the monkey, feet in the dog or cat, hoofs in the cow or horse.

Special Anatomical Features:

Ear.—The ear is organized purposely to make hearing possible. With its ear-drum, chain of bones, and cochlea, it is a very complicated device. In the cochlea are 24,000 fibers that respond to sound waves, and give us an impression of the pitch and quality of the sound.

"He that hath ears to hear, let him hear." In the Bible the ear is used as a symbol of that quality of the mind that receives and accepts the words of truth. "Itching ears" represent the desire to hear something new, to never be satisfied without hearing fables, speculations, and gossip.

Eye.—Research in physiology proves that the eye has 300

devices, all of which are necessary to sight. These are fitted together to make what is probably the most marvelous mechanism known. It is designed to make sight possible yet it develops in darkness.

In the Bible the eye represents spiritual understanding. Matt. 6:22. Wicked men are blind, but the righteous are able to see the truth. "Open thou mine eyes, that I may behold wondrous things out of thy law." Ps. 119:18. Darkness of soul comes to one who shuts out the light of love and truth.

Many nocturnal animals have eyes that glow when light is shone on them. This is due to a special reflecting device in the eye, that enables the animal to make use of all the light-rays available. It is present in moths, frogs, snakes, cats, dogs, raccoons, and many other night animals.

Foot.—The human foot is designed for walking, but not for running, crawling, leaping, or climbing. The two arches of the foot give a springiness to the step that is not found in any other creature. God made man upright, and fitted him for an upright life.

The horse, with his single hard hoof, is fitted to live on hard ground of the plains. Cattle, with split hoofs, can walk on much softer ground, in wet meadows where horses would be mired.

The caribou's hoof is very wide, so that it does not sink into the snow. During the winter the inner part shrinks, leaving a sharp edge that prevents slipping on the ice.

Hand.—The human hand is a masterpiece of mechanical design. Its flexibility, its position, and its gracefulness, make it more wonderful than any other organ. But even more wonderful is the way in which the hand is connected with the brain, so that mechanical skills are possible. The delicate operations performed by the hand are the result of an exceedingly intricate neuro-muscular mechanism not found in any other living creature.

Skin.—The skin of various animals is provided with means of cleansing. Frogs produce a mucus which carries away dirt. Outer layers of the skin are shed, sometimes in small quantities, sometimes all at once, as in a snake.

The heavy layers of fat beneath the skin of warm-blooded sea animals are for protection against the icy waters.

Smell.—Night animals have a better sense of smell than animals that roam about by day. In this way they compensate for the lesser powers of vision in the night.

ANIMALS.—See also Anatomy; Crayfish; Reptiles.

Animals may teach us of the care of God. Job 12:7-9.

They gather their food from God. Ps. 104:14, 21, 27.

They will be tame in the New Earth. Isa. 11:6-8; 65:25.

They can be tamed more easily than the tongue. Jas. 3:7, 8.

Animals have degenerated since creation. Many huge kinds have become extinct; others have become corrupted. Sin has produced many undesirable results in animal life.

Animals cannot reason like man. An animal may be trained, but cannot develop character. A child must be trained, and taught to act from principle.

Cruelty to animals is offensive in the sight of God. Men should learn to be kind to them.

Each animal is fitted with certain qualities and attributes that make it useful in its place in nature. What use would a dog be if it had a body like a hog, or the disposition of a sheep? Of what value would a horse be if he had the temper of a tiger? Or suppose a tiger had the disposition of a horse, how could we use him? Had a sheep been covered with bristles instead of wool, of what use would it be? Or if a pack of wolves had been clothed with wool, how would we obtain it? Suppose sheep gave milk and cows were covered with wool, would they be as good as they are now?

Lessons from specific animals:

Bats.—A bat may be blinded by covering its eyes, but will still be able to fly about a room filled with strings, without hitting them. This is possible because of thousands of sensory hairs over the surface of the wings. By means of these the bat is enabled to detect its insect prey in the twilight, when vision would be almost useless.

Bear.—The polar bear is especially adapted for its life. Its white coat makes it nearly invisible on the snow. The shape of the head is such that the nose, eyes, and ears are easily kept above the water when swimming. The feet are much larger than the feet of other bears, and there is webbing between the toes. The soles of the feet are covered with hair which prevents slipping on the ice.

Beaver.—Beavers show a high degree of intelligence. They gnaw through trees and use them for constructing dams and for food. They build canals to enable them to transport timber to their ponds.

The location of the house requires foresight. It must be in or over a pool deep enough not to be frozen to the bottom in winter. It must be so situated that it will not be destroyed by floods nor lack water in dry seasons.

In places where there are high banks, the beavers may construct long slides by which they travel from the water to the bluffs, and over which they transport supplies.

In gnawing a branch in two, cuts are made from both sides. Thus smaller cuts are needed than if they were gnawed clear through from one side.

Camel.—The camel is specially fitted for life on the deserts. It can go many days without water, as it has storage cells in its stomach. Its padded feet hold it in the soft sand. Stiff hairs guard the eyes. The nostrils can be closed to keep out the flying sand.

Cattle.—The stomach of cattle fits them for gathering large quantities of coarse herbage. It is swallowed only partially chewed, and stored in a large sac-like stomach. Then, when a sufficient amount has been gathered, the animal retires to a quiet spot and "ruminates," or chews the cud.

Cattle have a strong digestive juice that is capable of digesting cellulose, the woody part of grass, herbs, and branches of trees. Man cannot do this, but these animals are fitted to live on the coarser vegetation.

Cony.—"The conies are but a feeble folk, yet make they their houses in the rocks." Prov. 30:26.

One of the most helpless of creatures, the conies live among the huge boulders of the rock slides of the high mountains.

Conies always have a sentinel on guard to give warning of danger. At the least alarm they run to shelter.

Conies are so small and defenseless that they could not defend themselves against birds of prey or predatory animals. Wisely they do not expose themselves. They keep within the shelter of the protecting rocks. They are wiser than man, who exposes himself to temptation when he knows he is weak.

Fox.—Foxes show a great amount of cunning in their mode of hunting. They will often outwit dogs or hunters, and escape when it might appear impossible.

Frog.—In frogs the circulation is sluggish, and the pressure is low. There are several pairs of "lymph hearts" to pump the lymph back along the sides of the body to the heart again. Birds and mammals, with their high blood pressure, have none.

Tree-frogs have sticky glands on their feet, which enable them to cling to vertical surfaces. On the throat is a thin sac which may be expanded by puffing it out with air. This rests on the surface of the water when the frog is "singing," and adds tremendously to its vocal power.

Gopher.—Pocket gophers have short stubby tails, that

are covered with sensitive hairs. When feeding at the mouth of their burrows, they never leave the ground completely. Upon being frightened, they can back quickly into the burrow, using the sensitive tail to guide them.

Horse.—Horses must pull together to accomplish work. Teamwork is necessary in any kind of mutual endeavor. It is as necessary in God's work as anywhere else.

Horses are used in the Bible as symbols of strength and of swiftness. God has endowed the horse with a high degree of intelligence. A horse can be trained to work and to perform many tricks. A horse is faithful to his master, and is a good example of how we should be faithful to our Master.

Horses have to be governed by bits and bridles, but may be trained to respond to the word of the driver. Man should learn to be governed by the will of his Master.

Kangaroo,—Kangaroo Rat.—The tail of the kangaroo is a balancing organ. In similar manner the long tail of the kangaroo rat and other jumping rats and mice maintain the balance when the animals are jumping.

Lion.—The lion is used in the Bible as a symbol of Christ. Gen. 49:9, 10; Rev. 5:5.

It is also a symbol of Satan, who, "as a roaring lion, walketh about seeking whom he may devour." 1 Peter 5:8.

The lion stalks his prey with great caution, taking it when it is unsuspecting. At other times he roars to frighten it. The devil creeps up on men when they are unaware of his presence, then frightens them by his roaring, to make them think their case is helpless. When they have no courage to flee, he seizes them.

Porcupine.—The porcupine is protected by a coat of quills, and does not fear any enemy. If we are protected by the sharp darts of truth, we need fear nothing.

Porcupine quills are used by Indians in making beautiful decorations for their garments. Even the most undesirable persons may have some useful qualities, if we can find them.

Rabbits.—There is a type of rabbit for each kind of habitat, as for example:

Jackrabbit, on the open plains.

Cottontail, on the open forest and edges of thickets.

Brush Rabbit, in the brush.

Pigmy Rabbit, in sagebrush, the smallest of rabbits.

Seal.—Seals have a valve by which they can close the nostrils when under water.

Shark.—Small fishes with sucking discs on their heads often attach themselves to the bodies of sharks. The small fishes are carried about by the sharks, and are protected from large fishes that would otherwise devour them.

In turn, the small fish may free itself from the shark at any time it pleases. In return for its free transportation and protection, it frees the body of the shark from parasites.

Sheep.—The Bible uses sheep as symbols,—

1. of Christ. Isa. 53:7; Acts 8:32.

2. of Children of God. Ps. 78:52; 79:13; Eze. 34:11-16; Matt. 25:31-34; 1 Pet. 2:25.

3. of lost sinners. Luke 15:1-7.

Sheep are generally thought of as stupid creatures, yet they illustrate the quality of meekness and quietness. They follow their master or leader without question, even tho he might lead them to death.

"He maketh me to lie down in green pastures." As the sheep satisfy their hunger and lie down to enjoy their quiet rest, so the one who feeds on the Word of God finds rest and peace for the soul.

Squirrel.—The flying squirrel is fitted for its nocturnal life by the possession of large, glowing eyes.

Toad.—The tongue is fastened in front, and projects backward. By jerking the lower jaw down, the tip of the tongue is thrown out, thus catching the fly or other insect that is crawling past. Toads never touch anything except moving creatures. The tip of the tongue is sticky, and gets the unwary insect before it can escape.

Whale.—The whale is covered with thick hide loosened

up into innumerable open spaces which are filled with oily matter called blubber. This covering is two to three feet thick, and serves two purposes. It protects from cold (the whale is a warm-blooded animal), and it enables the whale to float on the surface.

Whales possess the same senses as land animals. They can see, hear, and smell. They sleep; they play; they are faithful to their mates; they manifest affection for their young. In these monsters God has planted the same qualities that He has given to other animals.

The lungs of the whale are located far back in the "hull," to assist in the balancing while under water.

The head of the whale has chambers in which air may be imprisoned while under water. The tubes leading to the lungs have plugs by which they are closed. When the air in the lungs becomes stale, the plugs are opened, and fresh air is taken from the chambers in the head.

Wolf.—In the scripture the wolf is used as a symbol of false prophets and false teachers. Eze. 22:23-28; Zeph. 3:1-5; Matt. 7:15-20; 10:16-18; Luke 10:1-3; John 10:11-13; Acts 20:28-30.

ANTS,—See Insects.

APHIDS,—See Insects.

APPLES,—See Fruit.

ART

"It is to Nature that Art turns for her finest inspirations, from Nature that she borrows her subtlest combination of form and color."—*W. H. D. Adams.*

"Can there be a grander and sublimer effect than that of the mountain peak which cleaves like a golden arrow the dim vast blue, or of the far-spreading plain of ocean with its ever-shifting lights and shadows."—*W. H. D. Adams.*

From sympathetic communion with the commonest natural objects, the soul may be trained to an excellent perception of taste. The graceful motions of the trees in a storm, or the curved lines of the retreating and advancing wave, as well as wild flowers or ferns, may impart deep impressions of grace and beauty.

All art is but the making of imperfect copies from nature. The great Master Artist is God. His beautiful master-pieces are everywhere.

ASP,—See Reptiles.

AUTUMN

The radiant glory of autumn is no less beautiful than the brilliance of spring. Over everything there is a spell that penetrates the soul with a mysterious power. So in life, there may be a sweetness in its closing days that is no less charming than that of youth.

In autumn the glow of nature, blazing in vivid colors on every tree and blade of grass, are like a consuming fire to cleanse the earth and make it sweet and clean for another year.

The leaves are dropped from all the trees, but they are converted into plant food, to be used again by God in building a new, fresh mantle of life for the next season.

Autumn warns that winter is coming. The wise man makes provision for the time of storm ahead. We need to heed the signs of the times.

> "And my heart is like a rhyme,
> With the yellow and the purple and the crimson
> Keeping time."—*Bliss Carman*.
> "The scarlet of the maples can shake me like a cry
> of bugles going by."—*Bliss Carman*.
> "The maple wears a gayer scarf,
> The field a scarlet gown."—*Emily Dickinson*.

BACTERIA

Bacteria serve many useful purposes. Many industrial processes depend on the action of bacteria to produce some chemical change. The characteristic flavors of many foods depend on bacterial products. In the digestive tract certain coarse foods that are not attacked by the digestive juices are broken down by bacterial action.

In the nodules of legumes are nitrogen-fixing bacteria that take atmospheric nitrogen and combine it with oxygen, thus "fixing" it, so it can be used by plants.

In the soil nitrifying bacteria take the ammonia resulting from decay of proteins and turn it into nitrates, in which form it can be used by plants. Diseases are caused by new strains or varieties of bacteria that are more virulent than usual, or by new varieties to which the body is not immune, or by the lowering of body resistance to bacteria already present.

When new kinds of bacteria are introduced into the body, there is immediately set up a production of chemicals known as antitoxins. These destroy the poisons, or toxins, thrown out by the bacteria. The presence of the antitoxins gives *immunity* against the disease, sometimes for a short time, sometimes for life.

Sin is like a great sore, or infection, in the life. Our resistance has been lowered until all kinds of germs of evil find lodgement within us. Of ourselves we have no means of resistance, but Christ acts as the great antitoxin to counteract the poison of sin. Not only that, but He enables us to build up a resistance to sin, until we become immune to its attacks.

BARLEY

In many lands barley is regarded as a symbol of chastity. The threatening arms stand out as if to say, "Touch me not."

BAT,—See Animals.

BEAR,—See Animals.

BEAUTY

God's beauty appears in nature. The "beauty of holiness" is His character. That holiness shines forth in myriad glories of earth, sea, and sky.

"He alone who recognizes in nature his Father's handiwork, who in the richness and beauty of the earth reads the Father's handwriting,—he alone learns from the things of nature their deepest lessons, and receives their highest ministry. In this book of nature opened to us,—in the beautiful scented flowers, with their varied and delicate coloring,— God gives us an unmistakable expression of His love. He has covered the earth with the beautiful green verdure, for He knew that this color would be grateful to our senses. Each beautiful thing in nature is a token of God's love and care. In the loveliness of the things of nature you may learn more of the wisdom of God than all the schoolmen know.

"The beauties of nature have a tongue that speaks to our senses without ceasing. And as we behold the beautiful and grand in nature, our affections go out after God."

"God, who created everything lovely and beautiful that the eye rests upon, is a lover of the beautiful. From the solemn roll of the deep-toned thunder and old ocean's ceaseless roar, nature's ten thousand voices speak His praise. In earth and sea and sky, with their marvelous tint and color, varying in gorgeous contrast or blended in harmony, we behold His glory."

"In itself the beauty of nature leads the soul away from sin and worldly attractions, and toward purity, peace, and God. There is a refining, subduing influence in nature. It is His design that we shall enjoy the charms of nature, which are of His own creating."—*Ellen G. White, Miscellaneous Selection.*

God is not content to make things useful; He makes them beautiful as well. He hides the machinery of living creatures under a perfect harmony of curving lines and

OF NATURE

blending colors. The framework of the earth is covered with grass and flowers and trees.

BEAVER,—See Animals.

BEES,—See Insects.

BEETLES,—See Insects.

BILLOWS,—See also Ocean; Sea; Waves.

The psalmist uses billows as symbols of overwhelming affliction. Ps. 42:7, 8.

BIRCH,—See Trees.

BIRDS,—See also Coloration; Feathers; Migration.

In Eden, the birds were not afraid. Examples of this state of affairs may be seen in our National Parks, where birds will come fearlessly around the camps. Only where they are hunted and destroyed do they become wild.

God cares for the tiniest sparrow that sings its song without fear. He protects them against Satan's attempts to destroy.

Children may join with the birds in songs of praise to God.

Birds sing their happy songs in spite of the fact that they are constantly exposed to danger.

When the nesting season draws near, the male birds establish themselves in certain territories, and move about from one singing post to another within the area. In this way they protect the boundaries of their territories.

The songs serve not only to mark the nesting territories, but also to attract the females when they arrive. Then, during the time while the eggs are being incubated, the male birds defend the territories against all intruders.

The volume of song diminishes as the young birds grow, and the task of feeding them increases.

It must not be thought that song serves only to mark the nesting territories. During the fall and winter male birds occasionally burst into sweet song, often much sweeter than that which they give during the nesting season. It is evident that they sing for sheer love of living.

The organ for producing song, the syrinx, is located at the lower end of the trachea, instead of at the upper end, as in man. By this arrangement the length of the trachea can be used as a resonant organ to reinforce the sound, and the throat can be used to modify the sounds. This is necessary if the song is to have richness of quality, inasmuch as birds do not have facial sinuses for resonance.

The nest-building always precedes the time of greatest abundance of food. The young are hatched in time to grow to maturity when the supply of insect food is at its height. The nesting season runs from January to August because different kinds of birds require different kinds of food.

In California the Anna Hummingbird begins its nesting in February. By that time there are several kinds of wild flowers from which nectar and insects may be obtained for food.

Christ used the fowls of the air as examples of trust in God. They sow not, neither gather into barns, and yet they are cared for.

Birds obey the laws of their life. They migrate from one climate to another, they build their nests and rear their young. In every activity they follow the particular life pattern of their kind.

Birds show many remarkable structural features that fit them for their peculiar mode of life. Only a few can be mentioned.

Birds often have extra long tracheae, which lie coiled under the skin or may extend into the body cavity. This arrangement allows them to extend their necks without pulling the lungs out by the roots. One mammal, a sloth, has a very stretchable neck, and is provided with a coiled trachea.

The whole body of a bird is filled with air. The bones are hollow; in the body are large air-sacs; the feathers enclose air-spaces. The air inside the body is heated ten to twenty degrees above the temperature of the human body, giving considerable buoyancy.

The air in the air-sacs in the bird's body may be shifted from one sac to another, so as to maintain the balance of the body during flight.

Birds cannot sweat. The air-sacs in the body assist in regulating the temperature.

When a bird is resting, it breathes by rib muscles the same as other animals. But when it flies, the rib muscles cease to act, since the powerful flying muscles must have a solid anchorage on a rigid bony frame. The wing muscles cause the air-sacs to expand and contract, and thus furnish an effective system for circulation of air while in flight.

Water birds are covered with a thick, oily skin and a thick coat of feathers through which water cannot penetrate. They float on the water like a cork.

A diving bird is able to expel the air from its body when it dives, thus making it heavier than when floating on the surface.

The feet of birds are adapted to their life. Land birds have short legs and heavy feet; wading birds have long legs; swimming birds have webbed feet; perching birds have slender legs and feet; scratching birds have stout feet and moderately long legs.

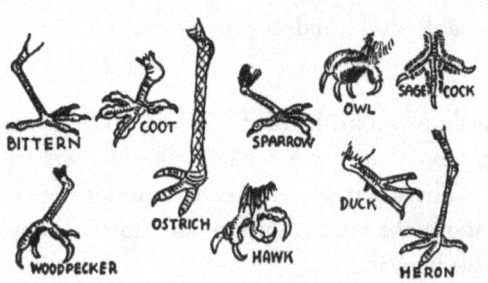

Grain-eating birds have crops in which they store their food, and from which it passes into the gizzard to be ground. Birds of prey have no gizzard, but a membranous stomach which secretes a strong gastric juice capable of digesting their food quickly.

Beaks of birds are fitted to the kind of food they eat. Seed-eaters have short, blunt beaks; woodpeckers have long, sharp beaks; insect-eating birds have slender beaks; ducks and geese have beaks fitted for gathering food from the mud and grass; hawks have hooked beaks. Many other types of beaks are to be noted, but each is exactly fitted to some particular type of food-getting.

Brilliantly colored birds are usually found in the open where enemies can readily be seen. Birds living in thickets are protectively colored, since they cannot so readily escape, and must depend on their colors to hide them.

Young birds build as perfect nests the first time as they do after years of experience.

Before a chick is hatched, a horny growth appears on top of the tip of the beak. This "egg torch" is used like a can-opener to release the chick from the shell. The same kind of structure is known in young snakes, lizards, crocodiles, turtles, and the egg-laying mammal *Ornithorhynchus* (duck-billed Platypus).

> "Books! 'tis a dull and endless strife:
> Come, hear the woodland linnet,

How sweet his music! on my life
There's more of wisdom in it."—*Wordsworth*.

Birds occupy what are known as "niches" in the natural scheme of things. Each kind has a separate and special place where it feeds. There is very little competition among different kinds. For example, the following food relations are to be seen in the forest:

Creepers fed on the bark, going up.
Nuthatches feed on the bark, going down.
Woodpeckers feed on the trunks and branches, digging in.
Chickadees feed on the smaller twigs.
Kinglets feed on the smaller twigs and foliage.
Warblers feed on the ends of the twigs and in the air.
Juncos feed on the ground.

The migrations of birds are guided by the hand of God. They obey the laws of their life better than man does.

Ships in the Mediterranean sometimes become the resting place of many migrating birds. Owls, hawks, and many kinds of small birds rest together on the rigging. In their common danger they forget their natural fears and animosities. So in face of common dangers men need to forget the fears and jealousies of life and live together in peace.

The arctic terns nest in the Far North, and migrate to the Antarctic, a distance of 11,000 miles.

The golden plover makes the longest flight of any bird, 2,400 miles at once, from Nova Scotia to South America.

> "He who, from zone to zone,
> Guides through the boundless sky thy certain flight,
> In the long way that I must tread alone,
> Will lead my steps aright."
> —*William Cullen Bryant, To a Waterfowl.*

Lessons from specific birds:

Bittern.—The bittern lives among tall grasses and reeds in marshy paces. It holds its long neck upright, and the stripes on the neck and body blend with the surrounding vegetation.

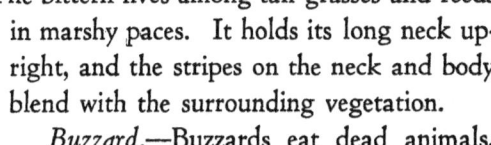

Buzzard.—Buzzards eat dead animals, and so do not have to catch them. They have hooked bills, but do not have talons like a hawk.

Coot.—Coots and grebes have flanged toes instead of webbed toes. This enables them to walk on mud flats as well as swim in water.

Cowbird.—The cowbird follows cattle, and as the cattle frighten up the insects, the bird takes advantage of it, and easily gets its food supply. In return, it rests on the back of the cattle, and by feeding on the flies that rest there, does a benefit to the cattle.

The cowbird does not find it convenient to have a permanent nest, since it is always following the cattle. So it has passed its nesting duties on to other birds. Its eggs are laid in other birds' nests, and they are left to rear its young.

Crane.—The crane is more obedient to God's will than are His people. Jer. 8:7.

Crocodile Bird.—The crocodile bird of Africa feeds on

the leeches that get into the mouth of the crocodile. The crocodile goes to the shore, opens its mouth, and holds the jaws open while the bird enters and obtains its food. Both it and the crocodile benefit from the arrangement.

Cross-bill.—The cross-bill at first appears to be deformed, as the upper and lower mandibles are crossed. However, it turns out that this is a highly effective device for prying open pine cones.

Dove.—The dove is used in the Bible as a symbol of the Holy Spirit. Luke 3:22; John 1:32.

Duck.—Ducks and geese have spoon bills with teeth along the margin. This arrangement enables them to strain food out of the mud.

Eagle.—The eagle represents self-exaltation. Obadiah 4.

The eagle deals with her young as God does with His children. She stirs up the nest, and forces them to strike out and fly. So God brings trouble and scatters His people to waken them from their slothfulness and love of ease. Deut. 32:11, 12; Ex. 19:4.

The line of swiftest descent of a falling body is not a straight line, but a *cycloid,* a certain kind of curve. An eagle descends in this curve when it sweeps on its prey.

Flamingo.—The flamingo has very long legs and neck. When it reaches down its head to obtain food from the mud, it holds it straight down, with the beak poining backward. This brings the

FLAMINGO AND NEST

opening of the bill in direct line with the surface of the mud.

Oriole.— "To hear an oriole sing
May be a common thing,
Or only a divine."
—*Emily Dickinson.*

Parrot.—Parrots can move the upper jaw separately from the skull. This is useful to them in using the jaws for clinging to a perch, or in obtaining their food.

Raven.—They do not store up food, but are dependent on God's care for a continual supply. Yet the raven lives longer than any other bird. Luke 12:24.

Before young ravens are fully feathered, the parents force them from the nest, and make them seek their own feeding ground. But God makes this instinct a means of feeding them. If they remained near the home nest, there would not be food enough, for all. By being driven from home, provision is made for them and their parents also.

Robin.—A robin finding an earthworm protruding from the ground will seize it and pull slowly until the worm is completely removed from its burrow. If the robin were to pull with a hard jerk, the worm would be broken. But under the steady pull, the worm relaxes its hold, and the bird gets it out whole.

Robins and other birds that feed on earthworms stamp on the ground to cause the worms to emerge from their burrows.

Sea-birds.—Gulls and other sea-birds will carry shell-fish up in the air and drop them on the rocks to break the shells and release the juicy morsels within.

Swallows and Swifts.—The swallows of the mission at San Juan Capistrano have left in the fall and arrived in the spring on the same days for a hundred fifty years, missing only two times in this period.

Chimney swifts, swallows, and whip-poor-wills feed on the wing, usually with the mouth open. They have wide, short, shallow bills, that enable them to easily catch insects in the air.

In swallows and swifts, which take their food entirely from the air, the wings are very long, while the feet are so small as to be almost useless. The mouth is large, and surrounded by hairs that assist in catching the insect food.

Thrasher.—The Le Conte Thrasher, nesting in the southwestern deserts in February, lines its nest with a layer of matted plant materials to keep out the cold.

Woodpecker.—The long tongue of the woodpecker is kept in a groove running around the side of the neck and up the back of the head. The muscles controlling the tongue are attached to the hyoid bones under the beak. These muscles pull the tongue down out of its groove and drive it forward, but cannot well return it. The return is accomplished by the springy action of the hyoid bones. The muscles driving the tongue forward straighten out the hyoid bones, which may back into place as soon as the muscles relax.

Woodpeckers have short, stiff feathers in the tail, which stick to the bark of the trees, and assist in holding them while drilling.

BITTERN,—See Birds.

BLOOD

Animals living in higher altitudes where the air is thin have an increased amount of hemoglobin. When a person goes into the higher altitude, his hemoglobin increases to compensate for the decreased amount of oxygen.

"The blood is the life." It carries the food to the tissues. It must be kept pure by correct habits of eating drinking. No poisons should be allowed to enter. The temple must not be defiled by tobacco, liquor, or harmful drugs.

BRAIN

The brain is the capital of the body. Correct physical habits are necessary to maintain good government (health).

Brain workers resist disease better than physical workers. The electric power of the brain vitalizes the whole body.

The brain becomes intoxicated by cheap reading as truly as by alcohol or drugs.

BRANCH

The Bible uses a branch as symbolic—

1. of Christ. Jer. 23:5; Zech. 3:8; 6:12.
2. of Lucifer, as an abominable branch. Isa. 14:12, 19.
3. of Christ's followers. John 15:1-8.
4. of the Church, as a beautiful and glorious branch. Isa. 4:2.

BRASS

An obstinate person has a brow of brass. Isa. 48:4.

Vain Christians are as sounding brass. 1 Cor. 13:1.

BREAD

The Bible is full of symbolic references to bread. Among the most interesting are these,—

1. Christ is the true bread. Luke 22:19; John 6:48-51; 1 Cor. 10:16, 17.
2. We are not to live by bread alone. Matt. 4:4.
3. It is a symbol of spiritual food. John 6:48-51.
4. It is a symbol of love. Eccl. 11:1.
5. Bread of deceit is first sweet, then distressing. Prov. 20:17.
6. Bread is a gift of God. Prov. 30:7-9; Matt. 6:11.

Unleavened bread contains no yeast; so the Christian is to have within him no germs of sin. (See Wine.)

There is more religion in a good loaf of bread than many realize.

Bread is food for the body; it must be eaten and assimilated. God's word is food for the soul; it must be eaten and assimilated if the soul would live.

BRIER

The brier represents evil men. It is a degenerate structure, whose thorns have come as the result of sin. God created the plant, but sin has made the thorns. Eze. 2:6; Isa. 9:18; 27:4; Micah 7:3-7; Heb. 6:7, 8.

BROOK

As referred to in the Bible, brooks are symbolic of,—

1. A well-spring of wisdom. Prov. 18:4.
2. Satisfaction in God. Ps. 42:1.
3. Deceitfulness. Job 6:15-17.

Brooks are as useful as rivers. Though small, they make rivers by uniting their forces. Small talents are as necessary as large ones, and when united, may produce great effects.

When its way is roughest its song is the sweetest, as it bubbles and ripples over the rocks.

Nothing can stop a brook from its ultimate purpose. Place a dam in its course, and it fills it and goes on its way again, leaving a sparkling lake behind.

"And yet how grand
Might life become, could all but understand
The thoughts that flow with brooks in every glade."

BUCKEYE,—See Trees.

BUDS

Buds are symbols of righteousness. Isa. 61:11.

Buds are formed in late summer, and lie dormant all winter, ready to burst forth in the spring. They are like those hidden powers of mind and spirit that lie waiting the wormth of sympathy and love to bring them forth to the fulfillment of their possibilities.

"It is God who brings the bud to bloom and the flower to fruit."

Some open at the first rays of light; others are so closely stuck together that there must be many days of warmth before they show any signs of opening. So with human hearts. Some are so cold that it seems impossible to reach them, while others open at the first touch of God's love.—*(Henry Ward Beecher.)*

BUTTERFLIES,—See Insects.

BUZZARD,—See Birds.

CADDIS,—See Insects.

CARIBOU,—See Animals.

CAMEL,—See Animals.

CAT,—See Animals.

CATFISH,—See Fishes.

CATTLE,—See Animals.

CEDAR,—See Trees.

CHAFF

Wicked men are like chaff, and will be separated from the righteous as chaff is blown away from the wheat. Job 21:17, 18; Ps. 1:4; 35:4, 5; Isa. 5:22-24; 41:14-16; Jer. 23:28; Hos. 13:2, 3.

CHALK

The chalk cliffs are made up of innumerable skeletons of ancient sea creatures, but each one is perfect as if it were the only one ever made. So life is made up of minute tasks, trifling duties, but each one should be as well done as if it were the only thing we ever had to do.—*(A. Maclaren.)*

CHAMELEON,—See Reptiles.

CHERRY,—See Fruit.

CHESTNUT,—See Trees.

CLAY

Our sinful condition is like miry clay. Ps. 40:2.

When dug from the ground, clay is soft, and may be molded into any desired form. Later, when it is baked, it retains the form in which it was molded. Youth is the molding period. The fires of life soon harden the character, and it remains in the form that was given it in youth. Once set, it can never be changed.

CLOUDS

Many beautiful lessons have been drawn from the clouds by Bible writers, as, for example:

1. Our sins are like a thick cloud, but God will blot them out. Isa. 44:22, 23.

2. Our goodness vanishes like a morning cloud. Hos. 6:4; 13:3.

3. A boaster is like a dry cloud. Prov. 25:14.

4. The wicked are like a cloud carried by a tempest. 2 Pet. 2:17; Jude 12.

5. Anger is like a cloud. Lam. 2:1.

6. Worldly prosperity is like a cloud. Job 30:15.

"And clouds arise and tempests blow
    By order from thy throne."—*Watts*.

By some mysterious power the water vapor in the air, as it condenses into clouds, collects in bunches rather than in solid masses. Thus there is produced a marvelous array of aerial scenery. If this were not true the slightest condensation of vapor in the upper air would cover the sky with a haze that would deepen to a gray, gloomy covering. The beauty of clouds at sunrise, at midday, and at sunset, is possible because of this wonderful arrangement that permits them to be broken up into fragments instead of remaining solid.

Vapors arise from the briny ocean and from dark swamps. By the wonder of the distilling process they leave all their impurities behind. In the clear blue of the sky they become transformed into cloud forms of marvelous beauty.

Dark clouds hide the sun, and bring a raw chill to our bodies. Sin is like a cloud that shuts off the warming rays of God's love.

When the sun rises higher in the heavens, it melts away the clouds and penetrates the thickest gloom. The love of God cannot be shut out, but will reach the darkest corners of the heart.

Travelers in the deserts welcome a cloud at noonday to shield them from the burning sun. So will Christ be a protection from the searing influence of Satan.

In the eastern Mediterranean a small cloud is of great significance. So suddenly do storms arise that mariners have little time to prepare for them. The apparently unimportant cloud on the horizon may grow rapidly into a raging tempest. Even so in life we never can tell to what proportions a seemingly insignificant act may grow.

Often in the high mountains, we may be in the midst of a heavy downpour. But if we climb upward, we may get above the storm and come out into the beautiful sunlight of the peaks. From this outlook the tops of the storm clouds present a most gorgeous sight, like a sea of white foam stretched out below.

COAL

A piece of coal absorbs all the light that falls upon it. That is what makes it look so black. Selfish people are like coal; they swallow up all the blessings that God sends them. But let the same carbon that forms coal be converted into a pure crystal diamond, and it will sparkle and shine, reflecting to others all the light that falls on it. God is looking for diamonds, not for coal.

COALS

Coals of fire are like the burning power of sin. Prov. 6:27-32.

Returning good for evil is like heaping coals of fire on the head. Prov. 25:21, 22; Rom. 12:20.

Coals represent the power to cleanse from sin. Isa. 6:5-7.

COLOR

Color, like sound, is produced by a graded scale of vibrations. The color vibrations, however, are of a different nature than those of sound, and are much shorter and vibrate much more rapidly. The slower vibrations that produce visible color are about one sixty-thousandth of an inch long. When these enter the eye, they stimulate nerve endings that transmit a sensation of red color to the brain. The shorter vibrations are only half as long, and produce an effect of violet. Between these are all the colors of the spectrum, or the rainbow colors.

Pure white light contains all the colors. Colored objects retain certain colors and reflect back others. Leaves are green because they absorb the red and violet rays and send back the green. God sends all the light. The color of the object depends on what is used. So in the spiritual realm; God sends men the pure white light of truth. Some men retain one portion, some another. The difference between them lies in their reaction to the light, not on the nature of the light shed on them.

The world would not be interesting if every object in nature were the same color. In human beings there are many different talents, many different activities. Each person draws from the "Father of lights" whatever he needs, and reflects his own color to the world. The same source supplies the light to all.

There are some invisible rays, the infra-red and the ultra-violet; but they are no less powerful because they are invisible. In the spiritual world are unseen forces, but they are as real as if we could see them.

COLORATION,—See also Birds; Feathers.

Color patterns often make a bird or animal inconspicuous. Green caterpillars feed on green plants; grouse and quail live among dead leaves; many birds are olive-green; bark moths are mottled gray; the toad has a warty brownish coat; green frogs resemble the scum in which they float; nymphs of dragonflies are mud-colored; brook-beetles' larvae are stone-colored. Thousands of such cases are known.

Many animals are protected from enemies by their colors. In some the color pattern is irregular, in stripes or spots, so

"Look at the wonderful and beautiful things of nature." (*Steps to Christ*, p. 9)

"In nature itself are messages of hope and comfort." (*Steps to Christ*, p. 10)

as to camouflage the animal. Most animals are dark above and light beneath. In birds and fishes this makes them less visible when seen from below against the light background as well as when seen from above.

Many animals turn white in winter. Among these are the snowshoe hare, weasel, and ptarmigan.

The chameleon changes its color so as to blend with its surroundings.

Frogs and other amphibians are light or dark according to the background.

In coral reefs green fishes are usually found in vegetation; blue fishes live above the bottom; gray ones are around the coral heads; banded fishes live in many situations.

CONY,—See Animals.

COOT,—See Birds.

COWBIRD,—See Birds.

CRANE,—See Birds.

CRAYFISH

The crayfish, and lobster, shows in a remarkable degree the adaptation of one part to another and all together for their habitat. There are two long pincer feet in front of the body, with large pincers. When the animal needs to escape its enemies it could not move forward rapidly, as these feet would be in the way. So it moves backward by a quick downward stroke of the abdomen. This allows the pincer feet to drag along.

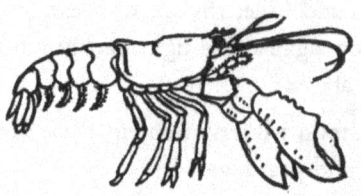

The gills are so arranged that a pair of gill bailers circulate the water from front to back through the gill chambers. Thus when the crayfish is swimming backward the circulation through the gills is assisted rather than hindered.

The green gland, or kidney, is located *in front of the mouth and gills*. The gill circulation carries the wastes away from the body. If the outlet from the kidneys were at the back end of the body as in most animals, the wastes would be carried to the gills.

These features all fit together to enable the animal to live efficiently in its watery home, both when crawling on the bottom and swimming through the water.

CREATION,—See also Nature; Nature Lessons.

The Sabbath was given as a memorial of creation. Ex. 20:11.

CROCODILE,—See Reptiles.

CROSS-BILL,—See Birds.

CRYSTALS,—See also Precious Stones.

"In the works of crystalization we behold the perfect figures of geometry, as traced by the finger of God."

CUTTLEFISH

The cuttlefish ejects a black fluid to hide it from its enemies.

OF NATURE

DARKNESS

Darkness is used throughout the Bible as a symbol of sin.

Many animals of prey prowl about in darkness. So men are said to love darkness rather than light, because their deeds are evil.

Yet darkness has its benefits. Some plants will not mature without several hours of rest each night. Human beings require a period of rest in sleep. So the darkness which in some cases is symbolic of evil, is in other cases a blessing. It depends on us as to the use we make of that which nature has provided for us.

DAWN

"Every dawn promises a day that will exceed it in loveliness, a shining more and more unto the perfect day."—*Marian M. Hay*.

DEER,—See Animals.

DESERT

Deserts did not exist in Eden. In the New Earth, they will be restored to their Eden beauty. Isa. 35:1, 2, 6, 7; 41:17-20; 43:18-20; 51:3.

In spite of its barrenness, the desert has its beauties. The play of color on the distant mountains, the sunrise and sunset, the soft blending colors of the landscape,—all have a charm not to be found elsewhere.

Desert air is pure and free from fog, smoke, and mist.

Yet, on the other hand, the desert can be very cruel. The traveler who is lost in the desert without water, may perish from thirst. Many besides the Israelites are wandering in the Desert of Sin.

Dust and sand storms arise in the desert, and may cover caravans that are overtaken by them. A traveler would be foolish to cross a desert without a guide or a good chart. We need a Divine Guide to direct us across the desert of sin.

The salt desert west of Great Salt Lake is covered with water at certain times of the year, and looks like a beautiful lake. Yet no life can exist in the briny waters. Satan may try to deceive us by covering his errors with some truth. But underneath is the salt and brine of wrong. It will sting if we get into it.

Juniors going to the camp at Idyllwild on Mt. San Jacinto in southern California have to ride for many miles across deserts and chaparral mountain slopes. Suddenly the car drops down a slight incline and arrives in a beautiful wooded valley. This same situation is found on the route to Sequoia, General Grant, and Yosemite national parks. The magnificence of the forested valleys is the more noticeable from the barrenness of the region through which one has just passed. The glories of heaven will be all the more appreciated after the trials of the last days.

DEW

Dew is like the life of a sinner: it soon passes away. Hos. 13:3.

Like dew in summer, prayer is silent. Though unseen, it produces immense results.

The smallest dewdrop has a star reflected from its shining surface, and the most insignificant passage of Scripture has in it a shining truth.

In the night, the dewdrops have been deposited on every leaf and blade of grass. When the sun comes up, they shine like diamonds. Formed in the darkness of night, they shine forth in beauty when touched by the sunlight. God's plans for us may be unknown to us, but they become beautiful when His purposes are revealed by the light of His love.

As the dew falls as willingly on a bit of dead wood or dry leaf as on anything else, the Christian should be as willing to occupy the lowliest place, if thereby he can accomplish some good.

The morning dew refreshes the flowers and fills the early morning with sparkling gems. In the quiet of the early morning our souls may be watered by the freshness of God's love.

In the hot valleys of California there is little dew in the summer. But in the higher mountain meadows there is formed each night a generous deposit of life-giving dew. So dense is the moisture that the lower layers of the atmosphere may be saturated, and every bush and tree will drip with condensed moisture. One must get above the lowlands in order to witness this refreshing bath that is given to the landscape each night.

No good deed is ever lost. From the ocean to the mountain snow, in rain or hail, or dew, as well as in the life stream of plants or animals, the tiny particles of water pass on and on, performing their mission of service.

"Every heart has lived through days of grief, and nights of loneliness and doubt, and God's promise to these is, 'My doctrine shall drop as the rain, my speech shall distill as the dew.' Deut. 32:2."—*Marian M. Hay.*

DIAMONDS,—See Precious Stones.

DISEASE,—See Bacteria.

DODDER,—See Plants.

DOG,—See Animals.

DOVE,—See Birds.

DROSS

The Scriptures liken dross to sin and wicked men. Ps. 119:119; Isa. 1:22; Eze. 22:18.

DUCK,—See Birds.

DUST

Dust is symbolic of humiliation. Gen. 3:14; Ps. 7:5; Isa. 26:5; 52:2.

Dust arises from the earth because of lack of moisture. Well-watered soil does not turn to dust. If a man's life is watered by the refreshing of God's spirit, there will not be the dustiness and desert barrenness that results from spiritual drouth. Some men's souls are as dry as the hills of Gilboa.

The "dust bowl" in the states just east of the Rocky Mountains has been the result of the removal of the natural cover of vegetation. God provided grass to hold the soil from drifting, but men wanted to raise wheat. They broke up the sod, and there was nothing left to prevent the dust

OF NATURE

from blowing. Man's interference with God's plans always brings trouble.

"Dust thou art." By His creative power God took the simple dust of the earth and formed the masterpiece of creation, a man in His own image. The transforming touch of God can turn even dust into a beautiful creation.

Dust in the air is responsible for many lovely sunsets and sunrises. That which we despise becomes the means of revealing God's glory to us. We should not underestimate the worth of the most lowly object of God's creation.

EAGLE,—See Birds.

EAR,—See Anatomy.

EARTH

There is a wonderful balance in the forces acting upon and within the earth. Its mass and weight are balanced with its distance from the sun and its speed of revolution so as to maintain its position in its orbit to absolute perfection. Gravitation on the surface is sufficient to hold the atmosphere in place, yet not so excessive as to make the pressure of that atmosphere oppressive.

EARTHWORM

Earthworms serve a useful purpose in digging up and loosening the soil so air and water can penetrate it more freely.

They make their homes by literally eating their way through the ground. The earth is moistened and passed through the digestive tube. In order to counteract the natural acidity of the soil, lime-producing glands pour a fluid into the stomach.

The earthworm breathes through its skin. Blood-vessels are scattered throughout the skin, which is very thin, and easily penetrated by oxygen.

EAST AND WEST

"As far as the east is from the west," is significant when we observe that there are no east and west poles. If a traveler goes in one of these directions he may go on forever without coming to the end. So, as our transgressions are removed from us, we can never find them again.

ECHO

An echo sends back the same words we speak. If we wish a good response, we must shout out a good word; if we receive evil words, it is our own fault. We get what we give.

An echo is always more pleasing than the original sound. So with God's answer to our prayers. He gives harmonious responses to our cries.

The echo does not always repeat the whole sentence. Only the softer and more pleasing words come back. Often God answers our shouting by only part of what we expect. He does not respond to the harsh calls we sometimes make.

Sometimes echoes pick up part of the voice and roll it back and forth like organ music. Heaven is able to transform our feeble efforts into divine melody.

OF NATURE

ECLIPSES

The moon, only 2,000 miles in diameter, can completely hide the sun, nearly a million miles in diameter, simply because it is closer at hand. Even a penny can hide the sun if held close enough. How often immediate needs hide the greater necessities, and blind us to the real essentials of life.

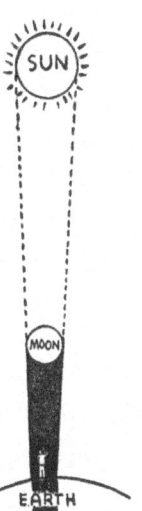

EEL,—See Fishes.

EEL GRASS,—See Plants.

ELECTRICITY

Electricity, conveyed to the seed, causes it to grow. So the power of God in the heart will cause it to grow.

When properly controlled, electricity is a valuable servant, but when out of control, it destroys life and property. In our lives there are many forces that must be controlled,—social, intellectual, political.

If we follow our natural inclinations, we may find ourselves at the mercy of uncontrollable forces that may bring death and destruction to us.

Electricity is invisible. We cannot see it, nor carry it about. Yet it is none the less real. By proper instruments it can be measured and directed as a servant of man. The spirit of God is invisible, yet when it comes into the life it becomes a powerful agent for good.

If the circuit through a lamp or through an electro-magnet is broken, it is useless. We must have the power of God continually flowing in us or we are powerless.

A slight film of corrosion over an electrical contact will often prevent the flow of current. Sin separates us from God and prevents His power from flowing through our lives.

ELEMENTS

God controls elements of nature and uses them as instruments to do his will.

Satan works through them to destroy.

All the marvelous variety of nature is built from about 90 chemical elements. The infinite power of God is able to take these few elements and combine them in endless combinations.

ELEPHANT,—See Animals.

ELM,—See Trees.

ENERGY

Energy is not inherent in nature. Nature acts only as power is supplied by God. "By Him all things consist," or hold together. He is the creator and upholder of all things. "Without Him was not anything made that was made." (Col. 1:17; Heb. 1:3; John 1:3.)

ENVIRONMENT

The adaptation of the environment to the needs of animal life is as remarkable as the adaptation of animals to the environment. Animal life is possible only within a limited range of conditions. The following physical factors must be properly balanced in order to make life possible: gravitation, pressure, temperature, light, sound waves, nature of the surroundings (whether solid, liquid, or gaseous).

The chemical factors, the composition of the substances about the animals, are also very important.

EVAPORATION,—See also Rain; Dew; Fog.

The fact that such a large amount of heat is required to evaporate water enables evaporation to serve as a highly efficient cooling agent. Sweat evaporated from the body, water from a lawn or field, rain from the surface of the ground or from a pavement, all carry away a large amount of heat. Thus the temperature is regulated by this wonderful property of water.

EVENING

"Men's lives should be like day,—more beautiful in the evening; or like summer,—aglow with promise; and like autumn—rich with golden sheaves, where good deeds have ripened in the field."

EVERGREENS,—See Trees.

EYE,—See Anatomy.

FEATHERS,—See also Birds; Coloration.

Feathers are a symbol of God's protection. Ps. 91:4.

Color in feathers is due to two causes,—sometimes to pigment, and sometimes to prismatic effects. The jewel-like throat of hummingbirds is caused by reflection of light from the inner structure of the feathers.

Patterns in feathers are due to many separate feathers that have grown into harmonious positions independent of each other. Darwin said the "eye" of a peacock's tail *made him actually sick* when he tried to explain it.

In many bright-colored birds, since the new feathers come in the fall, their tips are dull-colored, but wear off during the winter, thus revealing the brilliant colors in the spring, in the mating seasons.

Each feather consists of a shaft and a vane. The vane is made of barbs, and on the barbs are soft hair-like barbules. The barbules of one barb overlap or criss-cross those of the next. Where they cross are tiny hooks which cause them to interlock. Thus a soft, light structure is produced, having resistance, to the air.

Tail and wing feathers are stiff and heavy. *Contour* feathers cover the body, and the barbs are not hooked together. This makes them light and fluffy.

Beneath the feathers are smaller feather-like structures making up the *down*.

Woodpeckers have stiff hair-like feathers above the nostrils. These serve to keep dust and particles of wood out of the air passages.

A buzzard, when molting, drops feathers from both wings symmetrically. Thus the balance is more easily preserved than if one wing lost more feathers than the other.

Feathers are laid on the body shingle-like, from head to tail. This prevents the wind from blowing them up, thus cooling the body during flight, and also hindering its flight.

FERMENTATION,—See also Wine; Bread.

Passover wine was uncorrupted by fermentation, thus representing the sinlessness of Christ.

FIG,—See Fruit.

FIRE

The tongue is like a fire. Jas. 3:5, 6.

God's word is like a fire. Jer. 20:9.

Fire is one of nature's most beneficial agents. Plant substances used as fuel are so constructed that slow combustion is the general rule. It is impossible to release the energy in wood or coal or oil in such a manner as to bring about instantaneous burning. Only by man's artificial methods may products be made that will explode, or burn instantaneously.

FIRE-FLIES,—See Insects.

FISHES

The shape of the fish is the most efficient for traveling through the water. The front end is blunt, and the body tapers toward the back. This is the principle of streamlining which we see in modern planes and automobiles.

The eye of a fish is so shaped, and is built with a lens of such shape and proportion that make it of the proper construction for vision in the water. This same eye would be useless for sight in the air.

The eye of the *Anableps,* a fish of the rivers of eastern Asia, is divided into two halves, and each half is a perfect organ of vision. The lower half is near-sighted and the upper is far-sighted. Thus the animal is enabled with one pair of eyes (*bi-focals,* if you please), to see the little worms below that form its food, and with the other pair to see its enemies at a distance.

Along each side of a fish is a "lateral line," a row of sensory pits. By means of these the fish is able to keep headed up-stream in darkness or in muddy water.

The "swim-bladder" in fishes may be contracted to enable them to sink, or allowed to expand to lighten their weight, and alow them to rise in the water.

Lessons from specific fishes:

Catfish,—The catfish keeps her eggs in her mouth until they hatch.

Deep-Sea Fishes,—Deep-sea fishes are especially adapted in many ways. Some have extremely large eyes. Many have skin organs that produce light.

Electric Eel,—The electric eel has an electric storage battery. When touched, it sends out a strong shock which kills smaller animals and drives away possible enemies. In this way it obtains its food and protects itself.

Lungfish,—The African lungfish has skin glands that produce a varnish during the dry season when the fish is buried in the mud. This varnish protects it against drying.

Salmon,—Salmon are hatched at the heads of streams, and migrate to the ocean, where they spend several years. When they are full-grown, they return to the same stream where they were hatched, and there lay their eggs.

Stickleback,—The stickleback builds a nest to protect the eggs until they hatch.

FLIES,—See Insects.

FLAMINGO,—See Birds.

FLINT

Flint represents steadfastness. Isa. 50:7.

FLOOD

The temptations of Satan are like a flood of waters. 2 Sam. 22:5; Isa. 59:19; Rev. 12:15, 16.

FLOWERS,—See also Plants; Leaves.

Man's life fades as a flower. Ps. 103:15, 16; Isa. 28:1, 4; 40:6-8; James 1:11; 1 Pet. 1:24.

The flowers teach us of God's love and care. They carry the mind up to God.

"Every wayside blossom owes its being to the same power that set the worlds on high."

God would have the lives of the children beautiful with the graces of the flowers.

"Flowers are the sweetest things, that God ever made and forgot to put a soul into."

Flowers are the universal language of affection,—at marriages, sickbeds, funerals, to friends, at home, in church, everywhere. They bring the love of God to every one who sees them and smells their fragrance.

Some flowers are called "sunflowers" because they always turn their faces to the sun. They teach us to always turn our hearts toward the light of truth.

The sweetest flowers shed their perfume in quiet nooks, and the purest hearts perform their deeds of love in solitude and secret.

Cultivation of flowers brings children and youth into contact with nature and nature's God.

Roses and lilies and pinks should be gathered instead of thorns and thistles.

"The sweetness of divine love flows from His very presence as the fragrance from a flower."

Flowers are nature's poetry.

"The amen of nature is always a flower."

Wayside flowers are like Scripture truths. They are so common that we hardly notice them. Yet many of them are marvelously beautiful if we would stop to examine them.

"If there were no other proofs in the world of God's goodness, the flowers would supply them in abundance."

Alpine flowers bloom at the very edge of snow. Life is able to transform the waste brought down by the glaciers into gardens of beauty.

Thus we see how mercy and judgment are to be seen side by side.

"The flower fadeth"—Isa. 40:7. Unless it did so, the fruit could not follow. In our experience it may be necessary for some beautiful things to be taken away in order to make fruit-bearing possible.

The passing beauty of the flowers is like the fleeting pleasures of earth.

Lessons from specific flowers:

Lily,—God's care for the lily teaches us of His care for us. Matt. 6:28-34.

A lily is a symbol of a Christian. Hos. 14:5.

The lily springing from the black soil is brought to perfection by the power of God. "Even so will the life of God unfold in every human soul that will yield itself to the ministry of grace."

"Consider the lilies how they grow." We cannot of ourselves add anything to our own goodness. By receiving God's appointed agencies, we grow.

Lily-of-the-Valley,—The lily-of-the-valley is a symbol of Christ. Cant. 2:1.

Water-Lily,—The water-lily brings forth a blossom of beauty and purity even when it is growing in the slime of the muddy pool.

Orchid.—Orchids live in the dense jungles, where they are anchored to the decomposing bark and moss on the trees. From the foul atmosphere of the steaming jungle they draw nourishment, and transform it into gorgeous beauty. So with charity; it covers a multitude of sins, and breathes forth life, sympathy, and mercy.

Roses.—Choice roses are usually grafted on to wild roots. Sometimes the wild roots send up branches. These must be kept pruned off.

"If our faith is to grow high and bear rich clusters on

the topmost boughs that look up to the sky, we must keep the wild lower shoots nipped." (A. Maclaren.)

Severe pruning is necessary in order to produce the largest blossoms.

Sky Pilot,—The "sky pilot" is a *polemonium* growing in the Sierra Nevadas above 12,000 feet altitude. The glorious blue of these flowers seems to be a patch of the sky fallen to earth. One must climb the rugged steeps to obtain a view of them in their glory.

FOG

Fog is often beneficial. It prevents rapid evaporation. It shields the earth from the scorching rays of the sun. It allows much of the radiant energy of the sun to penetrate, but when the rays have been transformed to heat, they cannot pass back through the fog. So it blankets the earth.

Foggy weather is beneficial to young plants, especially those that have been transplanted.

The fogs of life sometimes appear to shut out the light of the sun, but when we understand their meaning, we often find that they have been a benefit to us. Christ told His disciples that there were many things that they could not understand. It is sometimes necessary for tender Christian plants to dwell in a partial light until they become strong enough to bear the full blaze of God's light of truth.

Fogs usually hang at low levels. To enjoy the view, one needs to climb above the mists in order to enjoy the purity of heaven's light.

Dense fogs hide the light of the sun, as moral darkness

hides the truth. As the light of the sun can scatter the fog, so the light of truth dispels moral darkness and gloom.

The fogs are only temporary. When standing on a mountain height we may find the landscape blotted from our view, yet we know that eventually the fogs will be driven away. Spiritual gloom cannot always exist; it will eventually be cleared.

FOOD

In nature elaborate food-claims become established. For example tiny green plants such as diatoms or desmids in a pond are eaten by bacteria, they in turn by protozoans, and they by water-fleas. The water-fleas furnish food for young trout, which might be eaten by larger fish or might grow to maturity.

FOOT,—See Anatomy.

FOREST

Down in the forest there is an "all-day service." It begins when the water of the little brook was lightened up with the golden rays of the rising sun. Prayer and praise rise continually from the violets and buttercups and daisies. Tiny rabbits look furtively on, and passing quail and doves lend their presence to the worship.

The elderberry on the edge of the swamp is not anxious to be the rose beside the stone wall. The lichen clings to the rock and does not disturb the moss by the creek. The oaks on the hillside do not try to crowd the alders from the side of the stream. Each stands where God planted it. If men were as satisfied with their allotted places, there would be no occasion for war and bloodshed.

If it was best for man in his original perfect state to dwell among the trees, how much more so for man in his restless state today, to lay himself open to the quieting influence of the forest. There is something in the woods that stills and sweetens the life, and restores the right balance of the mind.

> "One impulse from a vernal wood
>   May teach you more of man,
> Of moral evil and of good,
>   Than all the sages can."—*Wordsworth*.

FOUNTAINS

Empty fountains are like earthly wisdom.

"Cleanse the fountain and the streams will be pure. If the heart is right, your words, your dress, your acts, will be right."—*Ellen G. White*.

Christ is a fountain of truth. There is no possibility of exhausting the fountain.

Joseph was a fountain of life to Egypt. Joseph and Daniel poured forth blessings from God. So God's people are always to be.

FOXES,—See Animals.

FROGS,—See Animals.

FRUIT,—See also Plants; Trees.

Every tree is known by its own fruit. Matt. 7:16-20.
Good deeds are the fruit that Christ requires us to bear.

The fruits of an unselfish life will bear a harvest unto eternal life.

The apricot and the peach are the same; both have come from a wild tree of China. One has been selected for its juicy fruit, the other for its large seeds. They are excellent illustrations of the importance of choice. We may all have the same opportunities, but what we become eventually, depends on our choice.

"The perfect fruit of faith, meekness, and love often matures best amid clouds and darkness."—*Ellen G. White*.

Obedience is the fruit of faith.

It is not enough to be trees in the garden of God. We must bear fruit.

Lessons from specific fruits:

Apples,—(See also Oranges). Often when apples are in bloom, the codling moth lays its eggs in the heart of the flower. The fruit develops, and encloses the central portion where the eggs were laid. At the same time the worm grows in the heart of the apple, and when we gather the wormy fruit, there is nothing on the outside to indicate the presence of the worm within, but even though the fruit appears perfect it is defective.

Evil ideas are sometimes deposited in the hearts of children, and develop there all unknown to anyone. In later life, when the fruit is supposed to have been perfected, it may prove to be worthless because of some hidden evil that has been eating at the heart.

Cherry,—The Sierra cherry covers the bushes with beautiful red fruit, but no one will eat it, for it is very bitter. We cannot always judge by outward appearances.

Fig,—As the leaves of the fig were used by Adam and Eve as garments, so today men hide their sin beneath a robe of their own devising.

The barren fig tree was cursed. Christians who do not bear fruit will be cursed.

The tiny flowers of the fig are produced in great numbers in a hollow structure which on ripening becomes the fig. Commercial figs produce only pistillate flowers. Others, borne on other trees, are known as *caprifigs*. These contain both staminate and pistillate flowers. Upon entering a *caprifig*, the female fig wasp lays her eggs, and, unable to escape, dies there. The eggs hatch, and a new generation of wasps appears. When the females leave the *caprifig* they carry pollen with them. Those that enter the ordinary figs, crawl in, looking for a place to lay the eggs. They cannot find any place for the eggs, and die in the fig. But they have carried the pollen and dusted it on the stigmas, thereby stimulating the fig to grow.

Orange,—The word translated "apples" in the Scriptures, "like apple of gold in pictures of silver" should without question have been translated "Oranges." The fruit of the orange grows in a setting of silver blossoms. Both are on the tree at the same time. Such is a "word fitly spoken." No more beautiful picture could be imagined,—the golden fruit surrounded by pure silvery white flowers in a setting of glossy green leaves.

As the orange tree increases in age, the fruit improves in quality. On younger trees the fruit has thick rinds, but on older trees it is thin. So with man; as he grows older, his thoughts should become better, with less of the useless rind to hide their goodness from the world.

Plum,—Wild plums appear very beautiful, but may be so sour that they cannot be eaten. Not all that appears lovely is to be taken freely.

FUNGI

Fungous growth may be so small as to be invisible, yet in a short time it will destroy whole fields of grain. No sin is so small its presence is harmless.

In ancient times God sent "blasting and mildew" to punish Israel for their sins. He uses the commonest things as His servants for blessing and for punishment.

The spores of mildew, blight, rest, and other fungous diseases are everywhere by untold millions. The goodness of God restrains their growth, or all our crops would be destroyed.

Fungus grows best in darkness. As rays of light shine in upon it, its slime and mould is revealed. Some men's lives are such that it will not do to let the light of publicity shine upon them. They are the fungous growths of society.

Blight or mildew attack sickly plants more readily than

healthy ones. Only by maintaining a vigorous spiritual life can we remain unspotted by the blight and mildew of the world.

GALLS

When the gall-fly lays its eggs in the leaf or stem, a large growth takes place, due to the efforts of the plant to protect itself against the injury. This *gall,* as it is called, serves as a home, and food for the developing insect.

Galls are of great variety, both as to shape and color. Many of them are as beautiful as flowers. Thus by His wonderful power, God turns the injury into something beautiful.

GNATS,—See Insects.

GOD AND NATURE

He is the upholder of all things. Heb. 1:3.

God is seen in the works of nature, but nature is not God. He cannot be fully known by studying nature alone; His word must be studied also.

The flowers that adorn the earth are the spirit of God unfolding Him in visions of loveliness to our mortal eyes. Because of the force of His love, He must manifest Himself. Because of the beauty of His character, that manifestation must always be an enchantment.

"If Nature is not spiritual to you, it is because you are not spiritual. If you do not find the breath of Jesus in the balmy air, it is because His breath is too little in you."

"Nature has an utterly Divine mode of speech; her words are woven into delicious unities of form, into living

OF NATURE 61

trees and bushes, whose fire flows forth in roses and peach-bloom. Again she weaves her speech into singing larks and honey-bees."

"The beauty of all beautiful places says: 'God has been here.'"

GOLD,—See also Silver; Precious Stones.

Gold represents true character. Rev. 3:18; 1 Pet. 1:7.

Precious metal is refined by fire. Rev. 3:18.

"The gold tried in the fire is faith that works by love." The refining process is not easy.

A meek and quiet spirit is of more value than gold. Isa. 13:12.

GOPHER,—See Animals.

GRAIN,—See Seed; Wheat.

GRASS

The grass is symbolic of man's brief life. Ps. 90:5, 6; 92:7; 103:15; Isa. 40:6-8; James 1:11.

The Bible draws many lessons from the grass. Ps. 102:4; Prov. 27:24, 25; Matt. 6:30.

The face of the earth is beautiful by the abundant growth of green grass.

No matter how badly cropped or trodden, it springs into new growth.

The finest ribbon of man's manufacture cannot equal the richness and beauty of a blade of grass.

Waving panicles of grass carry a delicate beauty that is unsurpassed. Look at the individual flowers, and you will find in them the most exquisite delicacy imaginable. No artist can equal the lines that one finds in the grass flower.

Leaves of grass are spear-shaped, forming wedges that are easily forced through soil. When growing, they present little surface. Many can be packed together in small space without overcrowding and shading one another.

Many kinds of grass pour forth a rich perfume when cut. The odors of the hayfield remind us of people whose experience becomes richer under trial.

"Do not be troubled because you have not great virtues. God made a million spears of grass where He made one tree."—*Henry Ward Beecher*.

The Creator has repeated the same pattern more frequently in grass than in any other type of vegetation.

Grass is without branches. Wind gets little hold on them. They can be closely crowded.

"He maketh grass to grow upon the mountains." No matter how cold the winter or how dry and hot the summer, there will be a rich growth of grass as soon as moisture and warmth combine to make favorable conditions.

Grass is the most abundant and widely scattered of all vegetation. Perennial grasses are adapted to meadows that are always damp. Annual grasses grow where there are long hot summers.

Grass forms a sod to hold the soil from washing and the sand from drifting. If the earth were left barren, the top soil would soon drift away and other vegetation would suffer from severe drying.

The color of grass is due to a blending of the yellow and green rays that are reflected from it to the eye. The red and purple rays are utilized in photosynthesis, but the unused ones are the ones that give the landscape its beauty. For all we know, the Creator could have made chlorophyll so it would have used the red and yellow and green rays. In that case all vegetation would have been a rich blue and violet. The results would have been depressing. He might have used the green, blue, and violet. Then all nature would have shone with gorgeous red and orange color. This would have been too exciting and over-stimulating to the nerves. As it is, the green color is soothing and pleasing.

Even with green as the prevailing color there is endless color variation. There is a dark green and a light green chlorophyll, and mingled with the greens are soft shades of all the other colors.

"Teach me, Father, how to go
Softly as the grasses grow."—*Edwin Markham.*

GRASSHOPPERS,—See Insects.

GRAVITATION

Gravitation holds objects to the earth, and gives them weight. Occasionally we hear a person say, "How nice it would be if there were no gravitation. Then we could handle the greatest load with ease." This is a mistaken notion, for if there were no gravitation, the earth itself would fly apart, and nothing would be stable. The positions of the sun, moon and stars in the heavens depend on the action of gravitation. It is a universal law that holds everything together.

No one knows what gravitation is. The only way it can be measured is by its action on bodies. We call its effect *weight,* and measure it by a spring or a balance, in comparison to some standard we have chosen.

The Bible tells us that Christ holds all things together. "All things were made by Him." John 1:3. "By Him all things consist." Col. 1:17. "Consist" means to hold together.

Bury an acorn, and it will rise above gravitation, and become a mighty oak. Bury a stone, and it will always remain in the earth. The difference lies in the fact that there is life in the acorn. That life is able to overcome the pull of gravity. We need to have life within us, that we may rise above the force that would pull us downward.

GROUND,—See also Soil.

Ground represents the heart of man. Matt. 13:3-8; Luke 8:4-8.

Stony ground represents selfishness, where seed cannot take root. Matt. 13:3-6, 20, 21.

Good ground represents those who have the word and keep it. Satan cannot snatch the seed away. Matt. 13:8, 23.

HAIR

God's care for us is illustrated by the hair. "The very hairs of your head are all numbered." Matt. 10:30; Luke 12:7.

Many lessons in adaptation may be learned from study of hair, its structure and growth. Mammals are covered with hair, which serves in most cases as a covering for

OF NATURE

warmth. Hair and fur which is simply a thicker coat of fine hairs, serve also to protect from injury and wet. Hairs protect the eyes from dust, the ears and nose from crawling insects. Sensory hairs on the sides of the head and front legs assist many animals in detecting objects in darkness. Growth of hair contributes to the personal appearance of people and of animals.

HAND,—See Anatomy.

HARVEST,—See also Seed.

God's work in the world, and its culmination in the end of the world, is spoken of as a harvest. Jer. 8:20; Matt. 9:37, 38; 13:30, 39; John 4:35; Rev. 14:15. The harvest is multiplied by the seed sown in the soil. Without the sowing there would be no harvest. If in life we expect to gain anything worth while, we must sow the seed.

God exerts the same power in the harvest that was necessary to produce food for the feeding of the five thousand. It is manifested in a different manner, but is none the less miraculous.

The farmer sows in confidence; so are we to sow the seed, and wait patiently for the harvest.

HEATH,—See Plants.

HEN

The mother hen with her chicks was used by Christ to represent His care for His children. Matt. 23:37.

A sitting hen is said to be the most stupid and persistent creature one could imagine. Yet she illustrates the value of sticking to one thing until she gets results. For three weeks or more she can think of nothing else but her nest and eggs.

A hen with a brood of chickens is a match for any stray cat or dog. Her mother instinct makes her fearless. If we were as devoted to our responsibility as she, nothing could turn us from it.

HEREDITY

The principles of heredity are perhaps the most mysterious and marvelous of any known scientific laws. The hereditary qualities are carried in the cells of the body. In a cell 1/125 of an inch in diameter is a nucleus perhaps 1/500 of an inch in diameter. This nucleus holds, in a human cell, about 48 bodies known as chromosomes. Each gene is a factor which acts chemically on the cell. By the combined action of the genes the hereditary characteristics are controlled.

HERMIT CRABS

Hermit crabs are soft-bodied, and should be easily destroyed except for their habit of using cast-off shells for a home. When a hermit crab outgrows one shell, he searches for a larger one, into which he backs, with only his feet projecting. When he travels, he carries the shell with him.

HIBERNATION

In winter many wild creatures fall asleep in order to escape the shortage of food. Lying dormant throughout the winter, they are able to survive the long, cold winter when food is nearly impossible to find.

HONEY

Canaan was spoken of as a land flowing with milk and honey. Honey was thus symbolic of its prosperity.

Honey from different flowers is of different kinds. It varies in color and flavor. Some kinds are poisonous. The word of God is like good honey, variable in its nature. Some flowers of thought are polluted with poisonous ideas, and must be avoided like the poison honey.

Honey is preserved by the sting of the bee. Sometimes God places the sting of fear in His truth to give men caution in following it. We can not with impunity take upon ourselves the name of Christian, lightly regarding its obligations.

Some flowers pretend to have nectar when they have none. They are like false professors who make a show of religion, but have nothing to offer to others.

HORNED TOAD,—See Reptiles.

HORNS

Horns represent power. Ps. 75:10; Amos 6:13; Hab. 3:4.

HORSES,—See Animals.

ICE

Many beautiful lessons may be learned from ice. Its firmness and strength are in contrast with the weakness of water. It reflects the light that falls on it, having a clear bluish color. It stores up water for time of need.

If a piece of ice is struck with a hammer, it shatters, but if steady pressure is placed on it, it will yield.

Some untactful men try to drive others with hammer-like blows, and the result is a break of trust and confidence. The more gentle pressure would, undoubtedly, bring far better results.

When ice is cracked, it reflects rainbow colors. So often when our lives are broken by the hammer of God's discipline, we reflect the beautiful colors of faith.

Pound ice until it is shattered to powder, and it will freeze into a solid mass again. Warm it in the sun and it quickly melts.

You may argue with a man until you reduce his arguments to powder, but he will freeze solid again. Warm him with human kindness, and he will melt and become your friend.

ICEBERGS

In a storm at sea, the drift ice is battered and blown about, but the iceberg floats calmly on, unaffected by the wind and the waves. It is so deeply rooted in the water that nothing can upset its steady progress with the deeper currents.

Christians float in the same sea of life with other men, but they are so deeply imbedded in Christ that the storms of life cannot move them.

Some professed Christians are like icebergs. Their lack of cheerfulness is harmful.

"Nature is filled with spiritual lessons for mankind." —*E.G. White*

"Nature speaks to our senses without ceasing."
(Steps to Christ, p. 85)

An iceberg is mostly under water. True force of character comes from the degree to which we are anchored below the surface.

IMMUNITY,—See Bacteria.

INSECTS

The insects are one of the most wonderful groups of living creatures. In their almost endless variety of form and color, their adaptation to different conditions, and their usefulness and destructiveness, they teach us many important lessons. People from ancient times, have watched insects, and have used them as illustrations of truth. In addition to their value as object lessons they show us the marvelous power of God in providing such humble creatures with those qualities of body and those habits that make them successful against much larger animals and man.

The body of an insect is hollow, and filled with air-sacs, similar to those found in birds. Air-tubes extend throughout the body, and into the wings, where they form the veins. A hollow tube is known as the strongest construction possible for a certain weight. This tubular vein serves a double purpose, to stiffen the wing and to carry air to the tissues.

The compound eye of an insect is the most wonderful visual organ known. A dragon-fly has 13,500 facets, or separate eyes, in the compound eye. Each one is capable of perfect vision, yet when combined they give an efficiency of sight that is impossible in single eyes.

Insects are provided with some means of communication. Bees and ants will pass messages from one to another by means of their antennae. Presence of food or danger is quickly communicated to the whole swarm. Sentinels are kept posted at the entrance to their homes, to warn of danger.

Insects are far superior to larger animals in strength. Ants will carry fifty times their own weight. A beetle can move a hundred times its own weight. A house-fly beats its wings 600 times a second. A dragon-fly flies sixty miles an hour,—but it can stop instantly and go backwards or sidewise without changing the position of its body.

If a horse could leap as far in proportion to its weight as can a flea, it could leap over the Andes at one jump.

The powers of reproduction of insects are beyond comprehension. A termite queen will lay over two million eggs in a month. Thousands of insects are hatched in an hour and millions perish in a day. They are simply animated vegetation, as it were, specially prepared to supply the birds and other smaller creatures with food.

The colonial habits of insects are without parallel anywhere else in the animal kingdom. Bees, ants, termites, and others, construct remarkable homes, and have a complex social organization. (Their activities are worthy of careful study, but the subject is too extensive for this book. Look it up in an encyclopedia or book on natural history.)

Insects are ornamental as well as useful. Among them are found the most beautiful creatures in existence. This is true not only of butterflies, but of many other kinds. When examined under a microscope, they present a marvelous array of form and color. The Creator has adorned even the simplest of these creatures with grace and beauty.

Many insects produce irritating or disagreeable substances to protect them against enemies. The odor of "stink bugs," the woolly covering of woolly aphids, and the "elbow grease" that exudes from the leg joints of a blister beetle, are common examples.

Lessons from specific insects:

Ants,—Ants are object lessons of industry. Their homes are built by skill and perseverance, by carrying only one grain of sand at a time. They are a reproach to those who waste time in idleness. Prov. 6:6-8; 30:25.

Jungle ants will pick a skeleton clean of flesh quicker than a lion can. This is a good illustration of the results of cooperation.

An ant colony consists of parents, nurses, workers, warriors, slaves, architects, etc., each one helping the other.

The eggs will not develop unless nursed. The nurses lick the surface of the eggs. They carry them from one level to another to give them proper conditions of heat and moisture. When the larvae are hatched, the nurses feed them. When the larvae are grown, they spin cocoons. The nurses watch over the cocoons, and help the young ants emerge when they are mature. When the ants emerge from the cocoon, they are covered with a thin skin. The nurses pull this off carefully, and wash, brush, and feed the young ants.

Aphids,—Ants keep colonies of *aphids,* or plant lice,

from which they obtain a honey-like secretion. The ants move the aphids from one feeding ground to another. They carry the eggs of the aphids to their nests, and protect them through the winter.

Beetles,—Egyptians worshipped the scarab, or dung beetle, as a symbol of the resurrection. This beetle rolls up a ball of refuse in which it lays its egg, then buries the mass. In course of time the new beetles emerge from the corrupt mass, like a person raised from the dead.

The "water boatman" is a small aquatic beetle with large flat legs resembling oars.

Whirligigs are tiny water beetles that may be seen swimming in circles on the surface of a quiet pool. They have two sets of eyes, one on the under side of the head and one on the top. The lower set is for vision beneath the water; the upper is for vision in the air.

Bees,—Bees search deep in the flowers for the nectar.

Only bees with long tongues can reach the nectaries of certain lilies. The throat of these lilies is so small that ants and crawling insects are unable to enter.

Drones are not hairy, since they have no need of a hairy coat to collect pollen.

Bees do not mix different kinds of pollen together. Each one is stored separately.

A bee that gathers honey one day may gather pollen the next, but they do not mix their honey and pollen gathering.

Honey-comb is a remarkable piece of engineering construction. The cells are six-sided. This gives the greatest space, and the greatest strength with the least amount of material.

The end of the cell fitting into the septum, or division in the comb, is three-sided. Several bees work at once to produce this. When this three-sided pyramid is completed, the six-sided cells are constructed on top. Here again several bees work together, yet their work fits together perfectly. It is a marvelous example of how God teaches these creatures to do their work according to a perfect pattern, of which they themselves must be entirely ignorant.

The mason-bee closes the roof of its larval cell with a kind of cement, leaving only a little hole for feeding the young.

The carpenter bee makes a long tubular hole in wood, and divides it into chambers by cross partitions of saliva and sawdust. In each chamber it places an egg with a supply of pollen to feed the larva. In the first, or deepest cell, the bee bores a hole through which the young bee may escape. One by one as the larvae hatch, they bore through the partitions and make their way to the empty chambers, through which they escape. They never go in the wrong direction.

Mud-wasps fill their nests with spiders, which they paralyze by stinging. They do not kill the spiders, but sting them in a certain place, where the poison will reach the nerves in such a way as to deaden them but not kill. Thus the young wasps are supplied with live food when they hatch.

"Bees that have honey in their mouths have stings in their tails."—*Scottish Proverb*.

Butterflies,—Insects have long been regarded as symbols of the resurrection. The larvae of a butterfly or moth passes into an apparently dead pupa, from which it emerges as a beautiful creature.

Caddis,—The caddis-fly larvae construct tubular homes of sand and fine gravel cemented together and attached to stones in water. One species makes a silk net that it attaches to two stones, and strains out particles of food from the water.

Flies,—Flies are likened to folly. Eccl. 10:1.

Larvae of black flies are found in the surface of stones in very rapid water. A sucking disc at the back end of the body attaches them to the rocks. The mouth end is free, and has fan-shaped brushes for collecting food. Just back of the head are two suckers. By using these alternately, the larvae can move in the swiftest water. When it wishes to go down stream, it spins a thread, attaches it to a rock, and lets out the thread, or anchor-line.

Many flies have sensory organs which enable them to tell the direction of the wind. Thus they can poise in mid-air facing the air currents, and not be blown away.

Fireflies,—Some insects are luminous. The best-known of these is the firefly. By means of light flashes it illuminates its path.

Who can doubt also that this was meant to add beauty to the evening?

Gnats,—These represent unimportant matters. Matt. 23:24.

Yet how irritating a gnat may become, when it persists in troubling us. Some people persist in bringing up unimportant matters, to the irritation of those who are intent on more useful matters.

Grasshoppers,—Man's smallness is as grasshoppers. Num. 13:33; Isa. 40:22.

Locusts,—Swarms of locusts produce great damage in many countries. Even a small matter may produce great destruction if allowed to grow.

Mosquitoes.—Mosquitoes do not have eyes by which they can see at night, nor can they find their mates by smell. However, the female vibrates her antennae, which send out waves of extremely high pitch. The males have large feather-like antennae which pick up the waves.

How like a mosquito is the small, irritating influence that we cannot locate.

Moths,—Moths represent the destructive nature of sin. Job 13:28; Isa. 50:9; 51:8; Matt. 6:19, 20; James 5:2.

Flowers of yucca are pollinated by a small moth, the *Pronuba*. Pollen falling from the stamen cannot reach the stigma because the style hangs downward, and the stigma is hollow. The female moth visits one flower, collects some pollen in her mouth, and flies to another flower. There she lays her egg, then crawls down the style and pushes the wad of pollen into the hollow style. When the eggs hatch, the larvae feed on the seeds. Enough seeds are produced, however, to propagate the yucca. Without the moth no seeds would ever be produced, and without the seed the larvae of the moth would have no food.

Skipper,—The skipper, or water strider, is able to float on the top of the water. Its legs are so constructed that they support it on the surface.

Wasps,—(See also Bees.) Paper wasps build nests that are marvels of ingenuity. Woody fibers are gathered from old fence-posts, logs, etc., and are mixed with a sticky saliva and made into paper.

The organization of the wasp colony for defense is so effective that few persons care to interfere with their nests.

IVY

If the ivy be cut back to the ground, yet it will spring forth again. Not until sin is dug out by the roots will it cease to grow and bear fruit in our lives. Only the grace of Christ can destroy the roots.

We are like ivy that cannot grow or stand alone. Its beauty is seen only when it is hung up by a trellis or a wall. Our lives are beautiful only when we cling by faith to Christ.

JELLYFISH

The tentacles of the jellyfish have stinging cells. When they are touched, they throw out a poisonous fluid. Small animals are killed or entangled in the threads found by this fluid, and large ones are driven away. Thus the jellyfish is protected.

JEWELS,—See also Precious Stones.

God's children are like jewels. Mal. 3:17.

KANGAROO,—See Animals.

KANGAROO RAT,—See Animals.

LAKES

The shallowest lake may reflect the highest heaven. So the simplest mind may, when given to Christ, reflect His height rather than its own shallowness.

In Italy are two lakes lying side by side. But one is a light green color, while the other is dark and gloomy. One flows into the Adriatic, the other into the Black Sea. Their position side by side is no proof that their waters are bound for the same destination. They flow in opposite directions.

In California are two lakes, Tahoe and Mono. The first is one of the most famous lakes in the world. It is surrounded by forests amid which are nestled hundreds of cottages and resorts. The second is a briny dead sea inhabited only by brine shrimp and gulls. The difference? Tahoe discharges its water through the Truckee River to bless the deserts below. Mono gathers everything within itself and gives nothing away.

LEAVEN

Leaven, or yeast, is a symbol of sin. Luke 12:1; 1 Cor. 5:6-8.

The kingdom of heaven is like unto leaven. Matt. 13:33; Luke 13:20, 21. Hidden in the heart, the words of Christ permeate the life, and make a change in the conduct. They bring the mind and soul into harmony with the divine life.

LEAVES

The barren fig tree had nothing but leaves.

"Those who are but half converted, are as a tree whose boughs hang upon the side of truth, but whose roots, firmly bedded in the earth, strike out into the barren soil of the world. Jesus looks in vain for fruit upon its branches. He finds nothing but leaves."—*Ellen G. White.*

Leaf structure is adapted to the environment,—needle leaves to extreme cold or drouth,—dry hard leaves to semi-desert regions,—reduced leaves, as in cactus, to the desert,—large soft leaves to mild climates,—thick heavy leaves to tropical jungle.

Submerged leaves of water plants are finely divided. The reduced light is thus enabled to reach more leaf surface.

The pitcher plant lives where the soil is poor in nitrogen. In its leaves it traps insects, and from their bodies obtains its needed nitrogen.

In desert plants the stomata are set in deep recesses for the purpose of preventing evaporation.

Leaves fall in autumn to protect the tree; if they remained, they would drain the tree of moisture when the ground is frozen.

In California the tender-leaved evergreens have thick, heavy leaves that endure the cold. The madrone and manzanita shed their leaves in midsummer, when there is the greatest danger from drouth.

When a leaf falls, there is found a corky layer at the base of the petiole. This serves as a plug to stop the passage of water from the stem.

Fading leaves remind us of the end of life. "We all do fade as a leaf." Isa. 64:6. But even the falling of the leaves is attended by a gorgeous display of color. In the prospect of the end of life there should be hope, not dull brown despondency.

After the leaves have fallen, they contribute to the nourishment of other generations of leaves and of trees and flowers. A man's influence may count for good long after he has gone.

Leaves fade silently. It is not God's way to allow misfortune to be loudly proclaimed. He covers it with the tender beauty of love and sympathy as far as possible.

Old leaves clinging to a tree through the winter cannot be dislodged by the fiercest blasts. Yet in the spring they are pushed off by the new growth. So our old corruptions are best removed by the growth of new graces. It is the new life that drives away the sins of our old life.

Leaves take in the poisonous carbon dioxide and breathe out the healthful oxygen. So the Christian, dwelling in the poisonous miasmas of earth, should breathe forth the fragrance of life and truth.

Leaves buried in the rocks thousands of years ago have left their prints so accurately that they can be classified exactly. Every rib and vein is distinctly shown. None of us know how many people may read the prints of our lives long after we have gone. How important that our record be worthy!

On the under side of leaves are tiny pores, *stomata*, through which carbon dioxide is taken from the air and oxygen and water vapor passed out. On each side of the pore is a guard-cell. When the evaporation is so rapid that there is danger of the plant suffering from water loss, the guard cells shrivel and close the opening.

In leaves exposed to bright sunlight, the chloroplasts (chlorophyll-bodies) in the leaves are arranged in line with the direction of the light. Thus they help to shade one another, and reduce the heating effect of the light. In plants growing in the shade, and in heavy leaves that are in the sun, the chloroplasts are arranged horizontally, so that as many as possible are brought in contact with the light.

Wind often injures leaves. The canna and banana, with their long leaves, have made provision against this. The leaves may be slit transversely without affecting their efficiency. In the palms the large leaves are compound, so that slitting could not be brought about by wind.

LICHENS

Those who are always criticizing others are like lichens that fasten on the roughness of the rock. They are tough and hard and are dwarfed in growth.

Lichens are the first form of vegetation to appear when a region is made bare by flood or volcano or other devastation. Lowly though they are, they make it possible for other life to take root. They are the pioneers.

The simplest of plants, they are beautiful in form and color. Some are gray or blue or brown patches on trees or rocks. Some are yellow or green or pink. They brighten the landscape with their variety.

Lichens and mosses are most beautiful when the day is dark and cloudy. Then their soft colors show up to best advantage.

Nothing is lost in nature. Lichens serve to cover the bare rocks and convert their dullness into artistic beauty.

LIFE

Man's life is compared to a vapor. James 4:14.

As there is life in the seed, so there is life in God's word.

Every leaf pours forth life-giving oxygen by which men and animals live. Thus one life contributes to another, for plants in turn utilize the carbon dioxide that animals cast off.

The following quotations are from the writings of Ellen G. White:

"What science can explain the mystery of life?"

"In the natural world we are constantly surrounded by mysteries that we cannot fathom. The very humblest forms of life present a problem that the wisest of philosophers is powerless to explain."

"Life is mysterious and sacred. It is the manifestation of God himself, the source of all life."

"Through all created things thrills one pulse of life from the great heart of God."

"A mysterious life pervades all nature—a life that sustains the unnumbered worlds throughout immensity; that lives in the insect atom that floats in the summer breeze; that wings the flight of the swallow, and feeds the young ravens which cry; brings the bud to blossom, and the flower to fruit."

"The same power that upholds nature, is working also in man. The same great laws that guide alike the star and the atom, control human life. For all the objects of His creation the condition is the same,—a life sustained by receiving the life of God, a life exercised in harmony with the Creator's will."

"Who would dream of the possibilities of beauty in the rough brown bulb of the lily? But when the life of God, hidden therein, unfolds at His call in the rain and sunshine, men marvel at the vision of grace and loveliness. Even so will the life of God unfold in every human soul that will yield itself to the ministry of His grace, which, free as the rain and the sunshine, comes with its benediction to all. It is the word of God that creates the flowers, and the same Word will produce in you the Grace of His Spirit."

Nature makes almost unbelievable provision for the perpetuation of the species. Animals that face great dangers produce eggs and multiply in enormous numbers. Off the coast of Greenland the sea may be tinted brown with tiny jelly fishes, so small that a small glass could contain 3,000 of them. In certain parts of the Baltic Sea every drop of water contained 200 diatoms. A square yard of rock has been seen to have 120,000 mussels, and 200 young barnacles have been counted on a square inch of rock. An oyster may lay 50,000,000 eggs in a single summer. If all the descendants of a single green-fly should survive for one summer, the population would be greater than the human population of China. A single house-fly would become over a trillion in a summer if allowed to breed unmolested.

Thus life goes on! But why does not one species multiply until it fills the earth? Simply because of a system of balances by which the normal number of each kind is maintained. The surplus of each kind becomes food for some other creatures,—the smaller for the larger, until all the food chains are kept in operation, and life goes on its usual way.

Each blade of grass, each tiny insect, as well as the mighty tree and man himself, is a miracle of divine workmanship.

LIGHT

The Bible often refers to light in a symbolic way, for example:

1. The path of the just is as a light. Prov. 4:18.

2. The Word of God is a light. Ps. 119:105, 130; Prov. 6:23.

3. Righteousness is a light. John 3:19-21; Ps. 37:6.

4. God's people are a light. Matt. 5:14-16.

Light is a symbol of truth. As the earth is lighted with the sun, so it is to be lighted with God's truth.

As the lamps of the temple dispelled the gloom, so Christ will dispel spiritual darkness.

The first creative command was, "Let there be light." The first thing God does when he intends to recreate a man, is to turn on the light. As the light of truth shines into the heart, a new day dawns, and the work of a new creation has begun.

"If a thing reflects no light, it is black; if it reflects part of the rays, it is blue, or indigo, or red: but if it reflects them all, it is white. . . . This is the meaning of the 'white robe' which the saints wear in glory."

What the sunlight is to the plant, God is to the soul. He is the source of our spiritual life.

As the sunflower turns to the sun, we should turn towards God, who is our light.

Insects are attracted by bright lights, but are often killed by the heat. Not everything that attracts is safe for us to follow. Even though we may escape destruction, the wings of our spiritual experience may be so badly singed that we can never fly again.

"Light gives beauty, and reveals beauty already present. Plants could never develop their talent powers apart from the life giving light. No more can human beings expand and develop the possibilities of their souls apart from the Light of life."—*Marian M. Hay.*

LIGHTNING

In Ezekiel's vision the living creatures came and went like lightning. This was a symbol of the speed with which God's truth is to go to the world. Eze. 1:14.

Lightning is simply electrical power that has gathered

on the clouds until it bursts forth with a mighty flood of energy.

This same power may be harnessed and sent in a steady current over wires. Often human energies burst forth in uncontrollable force, and may do great damage. When harnessed and properly directed, this same power is capable of a great amount of good.

LILY,—See Flowers.

LION,—See Animals.

LIZARD,—See Reptiles.

LOBSTER,—See Crayfish.

LOCUSTS,—See Insects.

MAGNET

As the magnet points to the north, so the love of God points us to our duty.

Men are wanted who are as true to duty as the needle to the pole.

A needle touched by a magnet will never rest until it points to the north. Thus when our hearts are magnetized by the love of Christ, we never rest until we turn to Him.

A piece of iron will not attract others until it comes into contact with a magnet. We must come into contact with Christ before we can attract others to the truth.

MAN

Man is the most perfectly fitted for a highly intellectual and spiritual existence of any creature. The following features are worthy of careful study.

1. His upright position, with a head, large eyes and other sense organs, two arms and two legs, fit him for activity that is more efficient than any animal possesses.

2. His skin is a superior organ of sense, and protects his body against changes in temperature as well as against physical injury.

3. The human hand is the most perfect mechanical construction known. No other type of structure would make it possible for man to perform the large number of skilled operations that are done with the hand.

4. The bones and muscles are fitted together so as to give strength, ease of motion, and flexibility in a degree unknown in any other creature.

5. The brain and nervous system are vastly more complicated than in any other living thing. The mental attributes of man are superior to those of animals, and man alone possesses spiritual powers.

6. Man only has articulate speech, and the ability to form sounds into a language. Not only that, but he can write symbols that convey the same meaning as the sounds. Thus by spoken and written language culture is made possible.

MAPLE,—See Trees.

MEADOWS

Meadows may be occasionally flooded, but the marshes are drowned with every tide. Surprised by temptation, true saints are flooded with a passing outburst of sin: but the wicked live in it as their element. (Spurgeon.)

MICE,—See Animals.

MIGRATION,—See also Birds.

Tiny micro-organisms living in lakes and in the ocean undergo a daily up and down migration. They come to the upper levels when the light is dim and move to deeper water when the light is strong. Thus they keep themselves in the best light conditions.

MILDEW,—See Fungi.

MILK

Milk and honey symbolize prosperity. (See Honey.)

MIRAGE

"As waters that fail" Jer. 15:18. To the thirsty desert traveler the sight of distant water is a hope that spurs him on, often to leave him disappointed. The riches of this world are like a mirage that vanishes in thin air. We exhaust ourselves in pursuing it, only to fall fainting after our life has been spent in useless effort.

Mirages are like worldly pleasures. They leave us the more thirsty from the disappointment after they vanish.

MISTS,—See also Fog; Sun.

Mists represent obscurity. Acts 13:11; 2 Pet. 2:17.

The mists are dispelled by the rising sun. Even so the Sun of Righteousness scatters the troubles that hang over our souls.

MISTLETOE,—See Plants.

MOLE,—See Animals.

MOON

The moon is sometimes used to symbolize the church. As Christ is the sun, or greater light, so the church is the moon, or lesser light, that reflects Him to the world. Cant. 6:10; Rev. 12:1.

As the moon, though widely separated from the earth, travels with it, even so the church, while not attached to the earth, must go along to give light to the world as long as time shall last.

MORNING

Morning represents the Christian life. Job 11:17; Isa. 58:8.

It is also a symbol of the church. Cant. 6:10.

The morning of each day is the image of youth. The dew of God's grace lies fresh and cool, and the pure air and clear light shine anew.

MOSQUITOES,—See Insects.

MOSS

In California the moss on the trees becomes dry and appears dead throughout the long dry summer. But with the first rain it freshens and is as green and bright as ever. All that is required to bring the change is a bit of moisture. Many people are like the moss. They wither under the heat of criticism, but freshen with the moisture of kindness.

Moss, though lacking any true roots, is able to live on hard rocks and tree-trunks where higher plants would die. God has given to each plant and animal its place. Moss could not live in the cultivated field with the corn and potatoes. Men often fail to make a success in life because they try to live a life for which God has not fitted them.

MOTH,—See Insects.

MOUNTAINS
Mountains represent God's righteousness. Ps. 36:6.
They symbolize stability. Ps. 125:1, 2.
> "These granite rocks are organ keys
> His rivers play, and every breeze
> That whispers to the listening ear
> Sings in the anthem, "God is Here!"
> —*Rife Goodloe.*

"The mountains shall bring peace." Psalms 72:3.

As their summits rise into the clear sky, they stand as symbols. They are above the dust and smoke of the plains. They receive the pure light of heaven, stronger and clearer than do the valleys below. They are removed from the strife and turmoil of men. In the quiet of their deep canyons and lofty slopes rests a calm that brings rest and joy to the weary traveler who seeks their shelter for relaxation and recuperation.

In olden times men feared the mountains. Their rugged heights brought fear and terror. But now we appreciate that which was once feared. By learning to know and understand the beauty that God has placed in the mountains, we have come to look upon them in a different light. It is often so. We fear that which we do not understand.

Highlands and lowlands may fitly represent the ups and downs of our experience. We may live in the mountain of faith and righteousness or in the valley of sin and degradation.

Often the valley is cold and damp beneath a wet blanket of heavy fog, while on the mountain top the sun is warm and the air clear and bright.

Great effort is required to ascend a mountain. "The strength of the hills is His." We never realize the power of God until we match our puny strength against His mighty strength that piled the mountain on high.

The best views are always to be had from the highest peaks. On the mountain summit there comes an exhilaration of spirit that is worth all the effort to reach the top. So it is in any worth-while accomplishment in life.

The nearer we approach the summit, the smaller the mountain becomes, the narrower the path, and the steeper the climb. We must forsake the broad meadows, the rich forests, and finally the gravelly slopes. The last part of the ascent is over sharp, steep rocks. The nearer we approach to a worthy goal, the more effort is required to reach it.

The higher we ascend, the broader our vision becomes. Only narrow-minded, short-sighted men prefer to remain in the valleys of thought and spiritual vision.

Mountains serve a purpose that plains could never fulfill. They fill the thirst of the human heart for the beauty of God. Their rugged heights point to heaven, and their clear atmosphere is a symbol of that brighter, fairer land where the haze of earth shall be gone forever.

Mountain heights reveal their massiveness only when viewed from some equal height. The glory of Christ is appreciated only as we rise to the heights of Christian experience. As long as we are satisfied to remain in the valley, we shall never know the beauty of the mountains.

A mountain climber is never satisfied until he reaches the highest peak. Neither should we be satisfied with anything less than our best.

"Mountains hold commerce with God's upper ocean, and, like good men, draw supplies from the invisible."

"Mountain moss is always green."—*Henry Ward Beecher*.

Only from the top of a mountain can we see its general contour, and the relation of one part to another.

On the mountain top there is solitude. The higher a man rises, in spiritual and intellectual life, as in mountain climbing, the more isolated he becomes. But he has the joy of being above the world and nearer to heaven.

MUSHROOMS

Growing from decaying vegetable matter, the mushroom transforms corruption into beauty of form and color.

The brief life of a mushroom reminds us of the passing beauty of this world. Today it is a gorgeously painted structure; tomorrow it has faded away.

It is a fit symbol of the resurrection and the second death. Raised to life again, it soon passes away, never to be seen again.

Sometimes an apparently harmless mushroom is found to be fatally poison. Only a person who is on his guard can detect the difference between the good and the bad. But the failure to distinguish the poisonous kind does not save the person who eats it. So with evil. It counterfeits the good so closely that it is not easy to distinguish.

MUSIC

"There are wonderful cadences and modulations in the flow of the stream and the song of the birds, while indescribable harmonies are swollen by the myriad voices that go up from the eloquent earth."—*W. H. D. Adams*.

NATURAL LAW

Without natural laws, there would be chaos. Everything acts according to law. The stars, the sun, the planets, all living things, act in harmony with law. Heat, light, sound, electricity, mechanical processes, all are under law. Laws are possible only as there is an intelligent mind to formulate them. The laws of nature point to God as the ruler of the universe.

Natural laws are usually regarded as merciless. This is a false interpretation. Nature is full of healing power. No plant or animal exists but what serves to benefit others in some way. The scavengers of the earth help to keep it clean. Even the beasts of prey keep up the strength of the races of the kinds upon which they prey, by taking the weaklings and leaving the stronger ones to propagate.

NATURE

Nature has a ministry that is greater than most people realize. The following list suggests a few of the blessings to be obtained from a study of nature:

OF NATURE

1. It is an expression of God's thought.
2. It is God's character worked out.
3. It teaches us to obey God's laws. Its laws are certain; cause produces effect; there is no escaping the condemnation for wrongdoing. Briers and thistles teach us the lesson of the result of sin.
4. It bears the impress of the Deity.
5. All its blessings come as a gift of God.
6. Wherever we turn in nature, we hear the voice of God. It reveals His power.
7. It has charms that interest the child of God.
8. "Nature's ten thousand voices speak His praise."
9. It develops our spiritual powers. The most exalted truths may be brought home to the heart by a study of nature.
10. It is designed "to be an interpreter of the things of God." Harvests are as much a miracle as the feeding of the multitude.
11. Its lessons call for deep thought. Its mysteries teach us of the mysteries of God.
12. It is intended to make us happy.
13. It has a softening and refining influence.
14. It is the foundation for true education.
15. It teaches God's greatness. No two leaves are alike; no two sunsets are the same; one day is different from another one. No two persons were ever the same. In all nature there is infinite variety. Only the infinity of God's own character could produce this unending variety.

Nature is of itself incomplete; it leads the mind to seek for the fullness which it lacks. Spring leads into summer, summer into autumn, autumn to winter. Childhood looks forward to youth, youth to maturity. Each is beautiful in its way. None is complete. We look to God and the future life for a completion of that which is never fully satisfied here.

"Home can best be created in the country, and that is the reason, among many others, why children—circumstances permitting—should always be brought up there. The word "Home" then means to them something more than a place to eat and sleep in; their young minds are stored with recollections of all beautiful things in Nature, and of a thousand innocent amusements which the city child knows nothing of, although, by an undying instinct, he never ceases to pine for them. In the country, home strikes its roots deep into the heart; the children never forget the flowers they planted, the birds they were accustomed to watch, the little patches of garden which were given them to cultivate, the very sounds which are associated with the fields and woods. Some protection against evil is provided for the young when a love of this great world of nature is implanted in them."

"God has two gates by which He comes to us. Nature is His outer gate. He comes through the glories of earth and sky. Day by day, year by year, He reveals himself to us. Let us see Him, feel Him, in the wild flowers, the fragrant woods, and the happy birds."—*John Pulsford.*

"The good, the beautiful, and the musical are from one spirit. . . . Nature is the outsung poem of God's love, addressed equally to the eye, the ear, the smell, the taste, and the heart."—*John Pulsford*.

"Our capacity of appreciating the beauties of the earth we live on is, in truth, one of the civilized accomplishments which we all learn as an art."—*Wilkie Collins*.

"I cannot conceive of self-culture unless it includes the study of Nature, so as to render both the imagination and the intellect susceptible of the elevating and purifying inspirations."—*W. H. D. Adams*.

The beautiful is the shadow of God's majesty.
    "Nature is loved by what is best in us."
    "Waiting for worshippers to come to thee
      In thy great out-of-doors!"
              —*Henry Van Dyke*.
"God set seven signs upon this land of ours
  To teach, by awe, mankind His wondrous powers;—
  A river sweeping broadly to the sea;
  A cataract that thunders ceaselessly;
  A mountain peak that towers in heaven's face;
  A chasm deep-sunk toward the nether place;
  A lake that all the wide horizon fills;
  A pleasant vale set gem-like in the hills;
  And worthy younger brother of all these,
  The great Sequoia, king of all the trees."
              —*Charles Elmer Jenney*.
"I love not man the less, but nature more."—*Byron*.

"I know not where the white road runs, nor
what the blue hills are,
But man can have the sun for friend,
and for his guide a star."
>—*Gerald Gould.*

"I do not own an inch of land,
But all I see is mine,—
The orchards and the mowing-fields,
The lawns and gardens fine."
>—*Lucy Larcom.*

"There is a pleasure in the pathless woods,
There is a rapture on the lovely shore,
There is society where none intrude
By the deep sea, and music in its roar."
>—*Byron.*

"The Night is mother of the Day,
The Winter of the Spring,
And ever upon old Decay,
The greenest mosses cling.
Behind the cloud the starlight lurks,
Through showers the sunbeams fall;
For God, who loveth all His works,
Has left His hope with all."
>—*J. G. Whittier.*

"Somewhere primeval echo dies
Across the wastes untrod,
And wild and far and lone there lies
The wilderness of God."
>—*Harry T. Fee.*

"No character, I think, grows wholly ripe
Save that which grows as nature guides its growth."
—*Haydn.*

NATURE LESSONS

A sunbeam is invisible until it is reflected from some object. Divine truth is not understood until it is revealed in some natural phenomena. Nature is full of analogies to help us understand spiritual truth.

The tree of knowledge of good and evil was a symbol of the delusive range of temptation. Man was permitted to know that there was a possibility of temptation but he was instructed not to indulge, lest he die.

The tree of life represented the immortality that came from doing the will of God.

The following quotations are from Ellen G. White:

"Upon all created things is seen the impress of the Deity. Nature testifies of God. The susceptible mind, brought in contact with the miracle and mystery of the universe, can not but recognize the working of infinite power."

"To him who learns thus to interpret its teachings, all nature becomes illuminated; the world is a lesson book, life a school. The unity of man with nature and with God, the universal dominion of law, the results of transgression, can not fail of impressing the mind and molding the character."

"And as we behold the beautiful and grand in nature, our affections go out after God."

"From the solemn roll of the deep-toned thunder and old ocean's ceaseless roar, to the glad songs that make the forests vocal with melody, nature's ten thousand voices speak His praise."

"We may look up, through nature, to nature's God."

"In the briers, the thistles, the thorns, the tares, we may read the law of condemnation; but from the beauty of natural things, and from their wonderful adaptation to our needs and our happiness, we may learn that God still loves us, that His mercy is yet manifested to the world."

"Each beautiful thing in nature is a token of God's love and care."

"In earth and air and sky, with their marvelous tints and colors varying in gorgeous contrast or softly blended in harmony we behold His glory."

"God has revealed Himself to us in His Word and in the works of creation. Through the volume of Inspiration and the book of nature, we are to obtain a knowledge of God."

"We are to see and enjoy the works of God in the beauties of nature. Only he can fully appreciate the significance of hill and vale, river and sea, who looks upon them as an expression of the thought of God."

"All may find themes for study in the simple leaf of the forest tree, the spires of grass covering the earth with their green velvet carpet, the plants and flowers, the stately trees of the forest, the lofty mountains, the granite rocks, the restless ocean, the precious gems of light studding the heavens to make the night beautiful, the exhaustless riches of the sunlight, the solemn glories of the moon, the winter's cold, the summer's heat, the changing, recurring seasons, in perfect order and harmony."

"We are to see and enjoy the works of God in the beauties of nature. God calls upon teachers to behold the heavens, and to study His works in nature. The impress of Deity is seen upon the lofty mountains, the fruitful valleys, the broad, deep ocean. As we behold the beautiful and grand in nature, our affections go out after God. Wherever we turn, we may hear the voice of God, and see evidences of His goodness."

"The contemplation and study of God's character as revealed in His created works, will open fields of thought that will draw the mind away from low, debasing, enervating amusements."

NETTLES,—See Plants.

NIGHT

Plants and animals need periods of rest. Man wears his energies away with excitement and toil, scarcely stopping to rest. We need to take time to recover our wasted powers.

Even night is not a time of dead repose, for nature is building up her powers and preparing for another day.

"The darkness of night is a time of meditation, peace, and rest. There is too much that distracts in the broad glare of daylight. . . . But in the dim starlight the huge and thoughtful night draws near to soothe and calm and to fill the mind with great thoughts and feelings."—*Marian M. Hay.*

NOSE,—See Smell, under Anatomy.

OAK,—See Trees.

OCEAN,—See also Sea.

> "And the ocean's rhythmic pounding, with each
>   lucent wave resounding,
> Seems the music made when God's own hands
>   His mighty harpstrings sweep."
>
> —*Virginia Harrison.*

OIL

In the Bible, oil is symbolic of—

1. Joy. Isa. 61:3.
2. The spirit of God. Zech. 4:12; Matt. 25:3, 4, 8.
3. Love and kindness, as expressed in smooth words. Ps. 55:21; Prov. 5:3.

OLIVES,—See Trees.

ORANGES,—See Fruit.

ORCHIDS,—See Flowers.

ORIOLE,—See Birds.

OXIDATION,—See Fire.

PALM,—See Trees.

PARROT,—See Birds.

PEARLS,—See Precious Stones.

PINE,—See Trees.

PITCHER PLANT,—See Plants.

"The book of nature is a great lesson book..."
(*Christ's Object Lessons*, p. 24)

"He will unfold to us the beauty and glory of nature." —*E.G. White*

PLANETS

Each planet must have a certain speed of movement in accordance with its mass and distance from the sun, otherwise it would not maintain its position in its orbit. For example, Mercury has a speed three and a half times as great as Jupiter. Had Mercury been set moving with the speed of Jupiter, it would have fallen into the sun in a few days; and had Jupiter been given the high speed of Mercury, it would have gone off into space never to return.

PLANTS,—See also Flowers; Heath; Leaves; Pollination; Seeds; Trees.

Christ was like a tender plant. Isa. 53:1, 2.

"There's not a plant or flower below
But makes thy Glory known."—*Watts*.

Children of righteousness are as plants. Ps. 128:3; 144:12.

Evil plants will be destroyed. Matt. 15:13.

The growth of a plant from seed is like the growth of a child. They must be tenderly cared for and nourished.

Christian development is like the growth of a plant.

Many labors must be bestowed on plants in order to make them grow properly. So with Christians; they will not grow unless given care.

Air-plants, such as tropical orchids, have a mass of tissue around the roots that absorbs water from the air. The root obtains its supply from this coating.

Some plants turn the edges of their leaves to the sun on hot days in order to reduce the area exposed to the heat rays.

Leaves are always arranged in a mosaic in order to allow a maximum of sunlight to reach them.

Desert plants are adapted for storage of water. In many of these, such as in the cacti, the leaves are reduced or entirely lacking, and the stem carries on the work of photosynthesis.

Plants climb by means of tendrils, or by twisting the stem about a support. The tips are sensitive, and wave about in search of an object to which they can cling. As soon as they come into contact with such an object, they begin to twine about it.

A climbing vine has been likened to a Christian clinging to Christ for support.

Some plants require long days in order to mature their fruit. These are adapted to high latitude. Others require short days. These are adapted to the tropics. Thus the whole earth is effectively clothed with vegetation.

The kingdom of heaven is like a plant. In its earlier stages it rooted firmly in the earth. Then came the period of rapid growth. The final development of blossoms and fruit is awaiting the refreshing influence of the Holy Spirit to bring the final culmination of its growth.

Lessons from specific plants:

Algae,—The Germans call the tiny green scums and other algae "die Wasserblute," the water-bloom. So fast does it multiply in some cases that streams are nearly choked with it. It may not appear to the naked eye as anything but a mass of scum, but under a microscope it is revealed as a world of surpassing loveliness. Floating plants of many shapes and shades may be seen, triangular, round, square, and in chains, strings, and cylindrical strands. Some of them branch like a tree, others resemble feathers, others seem to be a succession of tassels. A world of beauty unrealized by the casual observer is opened to him who has eyes to see.

The common green scum of stagnant pools is so filled with oxygen on a sunny day that it floats by means of the bubbles of the gas that it liberates. Taking the deadly carbon dioxide exuded by the decaying matter of the pools, it transforms it into the life giving oxygen that men and animals require for their life.

Dodder,—The dodder has no roots of its own. It lives by sucking the sap from the roots and stems of other plants. It has no green leaves of its own. It produces no food. It is a parasite.

Did you ever hear of a "dodderer?" One who wastes his time, and lives at the expense of others, is no better than this parasitic plant. He is a dodder on society.

Eel Grass,—Eel grass grows submerged in large ponds and in lakes. The flowers grow at water level. In order to make adjustment for changes in level, the stem is coiled like a spring.

Heath,—Heath is a symbol of wicked men. Jer. 17:6.

The heath, or heather, is a small wild brushy growth outside the cultivated lands of England, Scotland, and northern Europe. It is similar in nature to what is known as brush, or chaparral, in different parts of America. The "heath" of Palestine is more like the chaparral of California. It is a stunted, dry, thorny growth, and was therefore chosen by the prophet Jeremiah to represent wicked men.

The word "heather" orginally meant a dweller in the heath, one who was outside the town. Symbolically, therefore, it is one who is outside the city of God, one who does not dwell with Him.

Heath may be cultivated, and may become very beautiful. So the person whose life is stunted by departure from God may develop a beautiful character if he allows himself to respond to the influence God tries to place around him.

Some heather plants bear most lovely blossoms. Even the roughest men have some good qualities, if we can but find them, and give them a chance to develop.

Many unpromising children may be changed by kindness, and blossom forth in beautiful traits of character.

Mistletoe,—Mistletoe attaches itself to the oak tree and draws from its supply of sap to carry on its own growth. In return it gives nothing. It is a purely one-sided, selfish arrangement.

Nettle,—Seize a nettle firmly and it will not sting you. Dangers are to be faced bravely, not dallied with. By meeting them with aggressiveness, we take away their power to harm us.

The flower of a nettle is harmless. Even the most dangerous situation may hold some good for us if we learn how to meet it.

Some people are like nettles. They sting whenever they are touched.

The poison hairs of the nettle prevent crawling insects from eating the leaves. The flowers, having no poison, are open to the butterfly and bee who will carry pollen and thus assist in pollenizing the flowers.

"When your tongue is ready with a sharp word to sting the offending person, seek to bring the nettle of your temper to blossom with the soft answer that turneth away wrath."—*Hugh Macmillan.*

Pitcher Plant,—See Leaves.

Poison Ivy—Hemlock—Oak,—The sumac family has several members that must be scrupulously avoided. A volatile oil given off by the stems and leaves, and contained in all parts of the plant, is extremely irritating to the skin of many people.

In the autumn these shrubs have the most gorgeous display of red leaves to be seen anywhere. Many an ignorant beholder has brought trouble on himself and others by picking them for bouquets.

And yet some grazing and browsing animals eat the leaves without injury, and certain small birds are fond of the berries in the winter. What is food for one is poison for another. We cannot all do everything alike.

Reeds,—A reed is used as a symbol of a weak sinner, who lacks strength to stand upright. Isa 42:3.

John the Baptist was contrasted to a reed shaken in the wind. Matt. 11:7.

Seaweeds,—Seaweeds are provided with "holdfasts" by which they cling tightly to the rocks. They are able to survive the fiercest waves.

The wonders of creative power are displayed in certain seaweeds growing in deep water. The red rays are unable to penetrate deeply, but some blue rays reach the deep water. Here the seaweeds are provided with a red pigment (it looks red because it absorbs the blue rays). Thus it can carry on its life with the power of the blue rays. If an ordinary green plant (which uses red rays) was placed there, it would die because it could not obtain the light necessary for its work.

Strangling Fig,—In the tropics is a vine known as the *strangling fig.* It attaches itself to a large hardwood tree and climbs on it. But it twines so closely that it destroys its host, and both fall to the ground together. So with anyone who climbs to success at the expense of others. They both suffer and the ultimate result is fatal to the parasite as well as to his victim.

Sundew, Venus Fly-trap,—The Venus fly-trap and the sundew are plants that grow in boggy soil where there is little nitrogen. When an insect comes in contact with the sensitive hairs on the surface of the leaf, the leaf folds up and encloses the insect. A gastric juice is formed, which digests the insect. When this is finished, the leaf straightens out again. In the sundew there is a drop of liquid on the leaf to attract the insect. This looks like dew, but does not dry away. For this reason it is called "sundew."

Vine,—The vine is a symbol of Christ. John 15:1-5.

It is also used as a symbol of the wicked. Rev. 14:18, 19.

Vines must have support. They represent our dependence on a higher Power.

All the branches draw sap from the same vine. We must draw our sustenance from Christ. Branches must be connected with the vine or they will die.

Unfruitful branches are pruned off.

Vines are fruitful on rough, barren soil where many other crops will not grow.

The vine is one of the most beautiful plants. Its beauty is quiet, delicate, and graceful. Perfection of form, color, light, and shade are found in a bunch of grapes. It is a fit symbol of Christ.

Grapes are the most fruitful of all plants. They begin to bear fruit at an early age and continue to a ripe old age.

The vine that trails along the ground is sure to be trodden upon. But if it will lift itself up from the earth and climb upon some firm support, it will not be bruised. Some people are always having their feelings stepped on. They need to climb out of the way of passing feet.

Violets.—All the culture of the world has been unable to improve the lowly violet. It is as sweet and lovely in its woodsy bank as in the most splendid garden. So some people are unchanged by circumstances. They are not disturbed by adversity or changed by good fortune. Their lives are true blue and fragrant in spite of their surrounding.

Wild Lettuce.—Wild lettuce has such thin skin that if an ant tries to climb the stem a white sap oozes out. This sap hardens and clings to the ant, preventing his further progress up the stem.

PLUM,—See Fruit.

POISON IVY,—OAK,—See Plants.

POLLINATION

Many beautiful lessons may be learned from a study of pollination. The flower furnishes nectar and pollen for the bees, and they in return carry the pollen from one flower to another.

Night-blooming flowers are white or yellow so as to be easily visible in darkness, and are provided with fragrant odors. Many of them are closed and odorless during the daytime.

Flowers usually droop during a rain in order to keep their pollen dry.

Wind-pollinated flowers do not have brightly colored corollas. The brilliant flowers are usually pollinated by insects. Insect-pollinated flowers are also equipped with odors to attract the visitors.

Sometimes the petals are equipped with guide lines (stripes) that direct the insect visitor to the nector pots.

These are always in such a position that the insect must pass by the anthers and the stigma.

In the iris the bee must pass the projecting stigma, and brush some pollen on it. After the bee has passed, the stigma springs back in place. His weight pulls down the anther, thus giving him a shower of pollen to carry to the next flower.

In mountain laurel, *Kalmia,* the anthers are held in pockets. When an insect enters, the anthers are released. The filament snaps upward, and the insect is showered with pollen.

Insects that crawl, such as ants, cannot help in pollination, and are smooth-bodied, with no hair to hold pollen. To keep them out, flowers either hang loosely or have a sticky substance or stiff hairs on the stem. Only a flying insect is allowed to enter the flower.

The milkweeds have their pollen in masses shaped like saddle-bags. The bee gets her feet tangled in the masses and pulls them loose, carrying them about for hours. Thus the pollen is scattered wherever she goes.

The *horsebalm* has four small petals and one large one. The bumblebee who is attracted by the strong odor, alights on the large petal and instantly slides off. Over her head hangs a mass of stamens. Grasping these for support, she becomes dusted with a shower of pollen.

The lady-slipper allows the visitor to enter; but once inside, she is trapped, for the entrance has closed. There is one way out, through a small opening at the back. In crawling through this, the bee must brush against the pistil and then against the stamens.

POOLS

The ugliest pool of muddy water is beautiful if we get at the right angle to catch the reflections from the grass, trees, sky, or clouds above it. Even the plainest lives may catch the beauties of nature and reflect them back. Men then see the beauty of the reflection, and exclaim, "What a lovely pool!"

PORCUPINE,—See Animals.

PRECIOUS STONES

Few things are more unlike than common clay and the precious rubies, sapphires, garnets, and carbuncles, yet the latter are made of the same elements as the clay. What an illustration of the power of God to transform common things into objects of beauty!

"As precious stones, polished after the similitude of a palace, God designs us to find a place in the heavenly temple."—*Ellen G. White.*

The New Jerusalem will be a glorious city, where shall be gathered together the brightness of the diamond, the ruddy glare of the topaz, the deep green of the emerald, the pure whiteness of onyx, the gold of the jasper, the blue of the sapphire, the violet of the amethyst, the iridescence of the opal, and the softness of the pearl.

God can make jewels out of the most worthless rubbish. Diamonds are merely crystallized carbon, the same substance that forms the soot in our chimneys.

No jewel is beautiful until it is cut. The polishing process requires constant holding against saws, grinding wheels, and polishing felts. So in human life, the roughness of human nature must be ground away before the beauty of true character can shine forth.

The Highlanders of Scotland search the mountain sides for sparkling rock-crystals that are washed out of the earth by violent showers. This is God's way of revealing His "jewels." Affliction lays them bare.

Lessons from specific stones:

Diamonds.—The effect of sin is graven on us as if made with a diamond. Jer. 17:1.

Diamonds are but crystallized carbon. The same material as the coal is used in fashioning the sparkling diamond.

A diamond is so hard that nothing can make a scratch upon it. Christians should be like diamonds. They should bear trial and hard treatment without being injured by it.

A diamond placed in water will shine, but a counterfeit appears dark and dull. The waters of affliction will not affect a Christian, but counterfeits will lose their appearance as soon as affliction covers them.

Pearls.—Pearls represent heavenly treasure. Matt. 13:45, 46.

Real pearls do not grow dim with age.

The pearl is begotten in pain. They are caused by the introduction of some irritating substance inside the shell of the oyster. The oyster proceeds to cover the irritation with layers of pearl. A single oyster that is ordinarily worth a fraction of a cent, thus becomes worth many dollars. Men whose lives might have been worthless have been ennobled and made glorious by some sorrow or hardship.

Precious truth is like pearls; it is not to be lightly regarded,—cast before swine.

A pearl glistens with rainbow colors, but a piece of ivory is plain in color. Both receive the same light, but the inner structure is different. Such is the law of nature and of spiritual truth. God's light is shed upon all, but the nature of the reflection depends on us.

QUICKSANDS

Quicksand appears as safe as any other, but beneath is wet, clinging earth, more dangerous than thin ice. Upon a touch the feet are drawn in, and to be drawn in is death. So with bad company. Externally it appears safe, but once we are drawn in, extrication is next to impossible.

RACCOON,—See Animals.

RAIN,—See also Dew; Fog; Mist; Storm; Vapor.

An oppressor is like a driving rain. Prov. 28:3.

Rain represents the Word of God. Isa. 55:10, 11.

It is a symbol of righteousness. Ps. 72:6; 2 Sam. 23:4; Hos. 6:3.

The early and latter rain are compared to God's power to be given to His people.

"He shall come down like rain upon the mown grass." Ps. 72:6. The influence of Christ's love is like the gentle rain that revives the fields after the hay has been cut. In a short time the dead brown stubble springs to life again.

"God comes down in the rain,
And the crop grows tall."—*Norman Gale*.

To change water from liquid to vapor requires a very large amount of heat. It requires 100 calories to raise a gram of water from the freezing point to boiling, but 538 calories are required to turn it to steam after the boiling point has been reached. When the vapor is condensed, this same amount of heat is released. Suppose now a mass of saturated air is struck by a wave of clear cold air. Condensation begins, and thus the heat stored in the vapor is released to warm the air. In this way the formation of rain modifies the effect of the cold wave.

Even though the condensation occurs below the freezing point and results in the formation of snow, the result is the same. The atmosphere is moderated.

Sometimes disappointment or trial have made our lives like a stubble field. God may have removed our joys and pleasures to make way for another crop that would be still better. But if we wait in patience, He will send the blessed rain of His presence to revive us again.

The blessings of rain are sent to all, to the wicked as well as to the righteous. God's mercy is extended to all. Matt. 5:45.

Tremendous power is wrapped up in small objects. One inch of rain over the United States would weight 45,000,-000,000 tons. All this is accomplished by tiny raindrops in such a quiet manner that it is hardly noticeable. God's ways are quiet ways. His blessings, both physical and spiritual, come quietly. His love descends like the shower of rain upon the earth.

Rain falls in drops, not in cataracts. Thus in gentleness it refreshes rather than beats to pieces the tender flowers. Falling slowly it gives the ground time to absorb its goodness and pass it on to the waiting plants before it is lost. God does not send His truth in torrents. Little by little He sends His word into our hearts, to refresh and revive us.

However cold and disagreeable rainy weather may be to us, it is necessary if we are to have fruitful seasons. So dark and gloomy experiences are sometimes necessary to prepare our hearts to appreciate prosperity and blessing when it comes.

RAINBOW

The rainbow symbolizes God's mercy. Gen. 9:12-17.

The bow in the cloud results from a union of sunshine and shower. So the bow over God's throne represents the union of His mercy and His justice.

In our lives the rainbow of God's promises is brought out by the sun of His love shining through the storms of our lives.

RAVEN,—See Birds.

REEDS,—See Plants.

REPTILES

Adder,—Wine stings like an adder. Prov. 23:32.

Evil men have adder's poison under their lips. Ps. 140:3.

Asps,—The heart of the wicked is like poison of asps. Job 20:14.

The poison of wine is like venom of asps. Deut. 32:33.

Chameleon,—If unable to reach an insect from one twig, the chameleon will try another perch. When finally he gets within "tongue-shot," he focuses both his eyes with their pinpoint openings on the victim. Rarely is a shot ever missed.

Crocodile,—The gavial of India is a large crocodile harmless to man. Old males develop a large tubercle on the end of the snout which serves as an air reservoir when they are under water.

Crocodiles, alligators, and some large lizards have a series of slat-like abdominal ribs to support and protect the heavy belly as it is dragged over the ground.

Crocodiles are able to close the back of the mouth off from the throat. Thus they can breathe with the nostrils above water, while still having the jaws open.

Horned Toad,—The horned toad is a lizard of the desert with a flattened, horny body. At night it lies in a depression that it has hollowed out of the sand. It is active in the daytime, when other animals find it necessary to seek a shelter. Many species will emit a stream of blood from the eyes when captured.

In some kinds of lizards the tail is snapped off when struck a sharp blow, or the animal may snap it off, leaving it wriggling before the astonished attacker while its owner makes his escape.

Serpents, Snakes,—The serpent is used in the Bible as a symbol of Satan. Gen. 3:3-13; Rev. 12:9; 20:2.

Wine is like a serpent. Prov. 23:29-32.

Christ called the Pharisees serpents and vipers. Matt. 23:33.

The jaws of snakes are attached to the skull by a movable instead of a solid joint as in most animals. This enables them to open their mouth and swallow prey that is larger than themselves.

Turtles,—In turtles the shoulder and hip bones are entirely inside the ribs which form the shell. This is a special adaptation to give them protection.

Turtles are fitted with water-sacs at the rear of the body. Water drawn into these sacs and expelled again, serves as a means of aerating the blood when under water where the lungs are useless.

Reptiles have a peculiar type of heart. There are two ventricles, but the wall between them is perforated. When a reptile, such as a turtle is under water for a long time, the lungs are inactive. The blood may then pass directly from one side of the heart to the other, without passing through the lungs.

RIVERS

In the Bible rivers are symbols,—

1. Of peace and joy. Isa. 48:18; 66:12; Ps. 36:8; 46:4, 5.
2. Of blessings. Isa. 41:17, 18.

As a river brings blessing to all along its banks, so the child of God brings the blessing of his religion to all with whom he comes in contact.

As the hillsides open a channel for the river, they are

repaid a hundredfold, so with the one who opens his heart to God's love.

"Even a human breast that may appear least spiritual in some aspects may still have the capability of reflecting on infinite heaven in its depths. . . . This dull river has a deep religion of its own—so, let us trust, has the dullest human soul, though perhaps unconsciously."—*Hawthorne*.

"I have lain all night a-listening
 To the voice of the water in the mountains,
Where in the white moonlight glistening
 Are assembled the mighty fountains."
—*Charles Elmer Jenney*.

ROBIN,—See Birds.

ROCKS,—See also Stones.

The smitten rock was a symbol of Christ.

Christ is the Rock of Ages, upon which the church is built. No other foundation will endure. Ps. 118:22; Matt. 21:42-44; 1 Pet. 2:4.

Rocks are used as a symbol of strength. A house founded on a rock will endure, all others will be swept away.

Many rocks are formed by the action of heat and pressure upon mud and sand laid down in water. These agencies transform the dull mud into rocks of beautiful color and texture. In life it often requires the heat of opposition and the pressure of difficulty to transform our dull lives into rocks fit for use in the Master's palace.

In many places in the rocks are found marks of raindrops that fell and were imprinted upon the mud, later to be hardened into rock. Even so good or evil leaves a print as clearly preserved as if it were graven in the rock.

ROOTS

Roots must run deep into the earth. Superficially rooted plants quickly die when drouth comes.

The drier the ground, the deeper the roots must go. Christ was like a root in dry ground. Isa. 53:2.

He went deep into truth. We must do the same, if we would withstand spiritual drouth.

Roots anchor the plant, and prevent it from drifting. We should be "rooted and grounded in the truth."

Roots often store food for a period of need. Thus they are enabled to bring forth vegetation and fruit under unfavorable conditions.

Desert plants have two root systems. The deeper system helps the plant to live during the long dry season. The surface system picks up the least bit of moisture that may fall during the rainy season.

Roots that meet with saline soil will push on through it, if possible, to find good soil below. When they find good soil, they send out many feeding branches to take in the nourishment.

When slips are placed in boxes of sand, there is a long period of time before any growth appears. They are making roots.

Students must spend a long time "making roots" before they are ready to take their place in life.

The Christian needs to "make roots" deep in Christ before he tries to bring forth fruit.

The root is a rough and unsightly part of a plant, but it is essential to life. Often God uses a rough, uncultured person in ways that are almost beyond our understanding. Many a humble worker, laboring at some menial task, is carrying on some part in connection with God's work that is just as essential as that of those whose labors make a larger show in public.

ROSES,—See Flowers.

SALMON,—See Fishes.

SALT

Salt represents wisdom. Col. 4:6.

It is also used as a symbol of righteousness. Matt. 5:13.

Salt preserves from corruption. We are to have salt in ourselves,—a saving knowledge of God.

The righteousness of Christ is as salt to preserve us from the influence of sin. It is a vital power in the life.

As salt mingles with other substances to save them, so the Christian must mingle with the world to save it. His influence must be diffused everywhere to prevent corruption.

SAND

Sand represents instability. Matt. 7:26, 27.

Solomon's wisdom was as the sand on the seashore. 1 Kings 4:29. Thus there is no limit to the wisdom of one who asks of God.

This wisdom does not come from a life of ease and luxury. Sand is produced by the action of wind and rain, waves and tide. By being battered and beaten, the rock fragments are finally reduced to the beautiful sand of the seashore.

Sand contains particles that have been brought from many original sources. Its beauty lies in the variety of color due to these varied sources. So our wisdom is of value as bits of knowledge have been gathered from far and wide. Our sources need to be broad and comprehensive.

Drifting sand, when driven by a hard wind, will cut away the hardest rock. Yet a soft, yielding substance is uninjured by it.

A wave that would break the hardest rock will fall harmless on the sand. The difference lies in the fact that the rock will resist, whereas the sand possesses the absorbing power of giving way to the pressure, and absorbing the shock. Some men are so stubborn that they resist any new ideas. Eventually they are broken by the impact of progress. Others are tolerant of the shock. They go on unharmed by attacks of opposition or criticism.

SAP,—See also Vine.

Sap is like the Holy Spirit, flowing through the vine, giving it life.

The flow of sap up giant trees like the redwoods is a mystery that man has been unable to solve. Just what forces can cause the sap to ascend 300 feet or more, is beyond the knowledge of any scientist. "Only God can make a tree," and only He can supply it with the living sap to fill its needs.

The evaporation of water from leaves assists the movement of sap up the stem. However, if a plant is placed in a tight chamber, droplets of water will appear on the leaves. Some inward force has driven the sap up the stem and into the leaves, even though there is no evaporation.

SAPLING

"As the twig is inclined, so the tree is bent." In early days the Indians used to bend down small saplings to make springs for rabbit snares. Occasionally one of these bent saplings would not be released and would grow in its bent position. Later in life it could not be straightened.

SCIENCE

Science is knowledge, and is usually thought of as knowledge of the natural world. True science teaches us the mighty power of God by revealing to us some of the ways in which He works.

Between true science and true religion there can be no conflict. Both reveal the same God. False science is like polluted water. The spring may have been pure, but man's theories and speculations have mingled the filth of the world with the stream of life.

Science that does not recognize God is like adulterated food. It is not the true manna that comes down from heaven.

SCORPION
Wicked men are like scorpions. Eze. 2:6.

SEA,—See also Waves.
The wicked are like a troubled sea. Isa. 57:20.

The wicked are like a turbulent sea, full of passionate desires, and never at rest; the righteous are like a landlocked lake, upon whose surface is reflected the beauty of the sky and the fleecy clouds.

In the darkest night, when storms beat upon the sea, there often radiates from each wave a soft phosphorescent light. So in life, the light of love illuminates the waves that beat upon us, and the gloom of discouragement is replaced by a soft glow of restful beauty.

Ungrateful men are like a salt sea, receiving the rain of heaven and the fresh streams from the mountains, and turning them all into salt.

The ocean gives back what it receives, in vapor that sends water back to the clouds; and so there is rain in the field and storm on the mountain, and greenness and beauty everywhere. *(Henry Ward Beecher.)*

Life is like sea-water; it never gets quite sweet until it is drawn up into heaven.

In the depth of the sea the water is still; the heaviest grief is borne in silence; the purest joy is unspeakable.

When the ship shakes, do not throw yourself into the sea. The ship rolls in the wind, but by the wind the ship advances.

The smallest brook can find the ocean; the lowliest child of God can find heaven if he keeps going in the right direction.

The waters of the earth are pulled up into tides twice a day, by the attraction of the sun and the moon. Tides affect the whole body of ocean, and near the shores become true translation waves. This causes a circulation of the waters that is much more efficient than could be produced in any other way.

Ocean currents are powerful agencies in maintaining a proper distribution of heat over the earth. For example, the mild climate of the Pacific Coast of North America is due to the warm Japan current, and the mild climate of the British Isles is due to the Gulf stream.

SEAL,—See Animals.

SEAWEED,—See Plants.

SEEDS,—See also Flowers; Plants; Trees.

Many beautiful lessons may be drawn from the seed. The Word of God is as a seed that is sown in the heart. Luke 8:11; 1 Pet. 1:23.

The germination of this seed represents the new birth. John 12:24; 1 Pet. 1:23.

The seed must be placed in the ground before it will grow. So man must give his life in service before he becomes of any value.

Germination of the seed is symbolic of the resurrection; it also is a fit symbol of the beginning of a Christian experience. "Whatsoever a man soweth, that also shall he reap." The seed is to be sown everywhere, and the harvest is sure.

The seed must not be planted too deep. Simplicity is needed. Many teachers and preachers sow so deep that the seed cannot grow.

On the other hand, it must not be left on the surface. Superficial work will not be of value.

Satan cultivates evil seed, and his harvest is seen on every side.

As there is life in the seed, so there is life in God's word.

The sacrifice of Christ is as seed sown in the world. The spread of the gospel is the result of this seed-sowing. There will be a great harvest in the kingdom of God.

Old neglected gardens that have lain untouched for years will yield a rank growth of plants of various kinds when plowed again. Seeds of good or evil may remain buried in the heart for a long time, only to spring into life with favorable circumstances.

SEED DISPERSAL

Almost an endless number of devices are used for seed dispersal. Briefly, here are a few:

1. Seeds provided with hooks or hooked hairs, such as cockleburs, beggar's ticks, burdock.

2. Sticky seeds.

3. Exploding capsules, such as snapdragons, violet, jewelweeds, and geranium.

4. Splitting or curling pods.

5. Edible fruit around the seed.

6. Wings or other devices for wind disposal.

7. Floating devices.

SERPENT,—See Reptiles.

SHARK,—See Animals.

SHEEP,—See Animals.

SHOWER,—See also Rain.

Showers represent the Spirit of God. Micah 5:7; Zech. 10:1.

SILVER

Silver must be refined to remove impurities. This represents the cleansing of the heart. Eze. 22:18-22; Mal. 3:3; Ps. 12:6; 66:10, 12; Zech. 13:9; Prov. 17:3.

The lost piece of silver represents sinners who are lost, but do not know it. Christ searches until He finds them.

Silver has always been prized for jewelry and plated ware because of its simple, plain beauty. It resists corrosion well, and is easily washed. It is a good example of the simple, quiet Christian virtues.

SKIN,—See Anatomy.

SKY

God produces scene after scene, picture after picture,

glory after glory, upon the continually shifting canvas of the sky.

"The great expanse of the sky has such vast and far-flung beauty that it is hard to think it exists for any other purpose than to surprise us with its unfailing variety and change."—*Marian M. Hay*.

> "The Almighty's mysteries to read
> In the large volume of the skies."
> —*Wm. Habington*.

> "The spacious firmament on high,
> With all the blue, ethereal sky,
> And spangled heav'ns, a shining frame,
> Their great Original proclaim."
>
> —*Addison*.

SMUDGE

In southern California the orange growers set smudges burning to keep off the frost. Garments, beds, faces,—everything becomes gray with a sticky, oily soot. If we live in an atmosphere laden with clinging evil, we shall become contaminated with its influence.

SNAKES,—See Reptiles.

SNOW

Snow is a symbol of purity. Ps. 51:7; Lam. 4:7; Isa. 1:18.

Snowflakes are exquisitely beautiful in design. Ice is

but a mass of flakes compacted together. The beauty is transformed into usefulness, for the ice locks up the water and gives it out slowly during the long hot summer. From the melting snows of the mountains we obtain our water supply and electrical power.

God brings order out of confusion and life out of death. The summer beauty of the hills and the fertility of the valleys, depend on the water wrapped up in the snow. The chill of hardship and the cold of trial may eventually result in beauty of character and in the development of useful lives.

Snow covers the brown and barren earth with a mantle of perfect purity, after the frosts have killed every green thing.

Sometimes our lives appear brown and sear from the chilling frosts of sin, but the forgiveness of God covers them with the snowy mantle of His righteousness.

Snow is a blanket to prevent loss of heat from the earth. In one Vermont winter, for four successive days, the air was thirteen degrees below zero. The soil under only four inches of snow, was nineteen degrees above zero,—a difference of thirty-two degrees. Thus the plants were protected from the intense cold.

Of the unnumbered billions of snow-flakes, few are ever seen of men. Yet each is as perfectly formed as if it were to be submitted to the gaze of the universe. Many people occupy a lowly place in life, and feel that there is no use in striving for the finer arts of living. Yet they may be as perfect and as lovely as if they were seen by all the world. Nothing truly beautiful is ever useless. Even if not seen by man, it is seen by God.

"As the snow gathers together so are our habits formed." One flake at a time produces the drift; one act at a time develops the character.

"God gives to the silent snow a voice, and clothes its innocence and weakness with a power like His own."

Soft, light, and beautiful as it is, it possesses tremendous power. In the blizzard the traveler is lost. Roads are hidden, towns are cut off from the world. Avalanches sweep down the mountain, leaving a wake of destruction in their path. Glaciers grind their rocky beds, and give rise to mighty rivers. Isa. 40:29. What a lesson on the power in little things.

> "But cheerily the chickadee
> Singeth to me on flume and tree;
> The snow sails round him as he sings,
> White as the down of angels' wings."
> —*Trowbridge.*

"And heavenly thought, as soft and white
As snow-flakes, on my soul alight."—*Trowbridge.*

SOIL,—See also Ground.

Labor must be expended in tilling the soil to make it bear. So in spiritual life; results cannot be expected without hard labor.

Soil is the reservoir from which plants draw their supply of nourishment. God has placed in the earth just the right balance of salts to make the growth of vegetation possible. Had the soil been strong in acids or alkalies, plants could not grow.

OF NATURE 129

SPARROWS,—See Birds.

SPIDERS

The hope of a hypocrite is like a spider's web. Job 8:13, 14.

The diving spiders construct a bell-shaped nest under the water. It then comes to the surface and obtains a bubble of air in the hair on its body. This it carries to the nest, and releases. The bubbles of air force the water out of the nest, and the spider lives in it and lays its eggs there.

The web of a spider is a remarkable piece of engineering. Many kinds of webs are constructed. The large orb web is the most commonly known. Some ground spiders form a tunnel in which they live, and a flat web to catch the prey. Others build a tunnel in the ground, and line it with silk.

Remarkable parental care is manifested by the spider that carries its eggs about in a silken case until the young are hatched. The garden spider weaves a silk cocoon about the eggs and suspends them in the web.

Balloon spiders climb up on fence posts or on the ends of branches and spin a tiny thread that is carried away by the breeze. When the thread is long enough, it bears the spider away, and sometimes carries it hundreds of miles.

A tiny strand of spider silk is used in a mighty telescope to enable the astronomer to measure the vast distances of the heavens. To what wonderful purpose is such a simple object thus applied!

SPRINGS

Christians are to be like a spring in the desert.

Christ is like a spring in the soul.

Springs burst from the earth without any visible source of supply. Year in and year out they continue to flow. Neither heat nor cold, storm nor drouth, can stop their flow. They have their source deep in the earth's water supply.

Jesus said to the woman at the well, "The water that I shall give him shall be in him a well of water springing up into everlasting life." John 4:14.

Artesian springs and wells gush up on the driest plain or desert. They are made possible by the presence of water-bearing gravel or porous rock that is connected with a higher area of sand, possibly hundreds of miles away. Without this constant source from the rains and snows falling on the distant highland, they would soon dry up. A spring is like living truth gushing forth from unknown depths. We cannot tell whence it comes, but it pours forth to freshen and invigorate us.

SPRINGTIME

Spring always follows winter. Nature never forgets. However long and dreary the winter, it is always brightened by the hope of a beautiful springtime.

As the earth blossoms forth into spring glory, so the heart renewed by the presence of Christ will develop anew.

In some lands it is hard to tell where winter ends and spring begins. But gradually one by one, the birds and flowers appear. With some Christians it is impossible to tell when the new birth took place. But the works of the Spirit declare that they have been revived.

Sin is stronger than winter, and evil habits are mightier bonds than crystals of ice; but God's power that brings the spring can break the bonds of sin.

SQUIRREL,—See Animals.

STARS

Stars represent angels. Job 38:7; Isa. 14:13.
Soul-winners shall shine as stars. Dan. 12:3.
The wicked are like wandering stars. Jude 13.

If a distant star should cease to exist, we would see its light for years, perhaps for centuries or millenniums. Even so, when a worker is gone, his influence continues to light the way for others.

The great planet Neptune was discovered because of its influence on the motions of Uranus. There are unseen influences in the universe, but they are none the less real. Some day, if we investigate, we may be led to recognize the existence of mighty truths of which we are entirely ignorant today.

Smoke and clouds may drift across the heavens, but after they have passed, the stars shine on. The goodness of God lives on in spite of obscuring circumstances. Some day it will shine out clearly again.

The stars have no bands or visible means of control, yet they never vary an instant in their courses. God's invisible power holds them. The same power can guide us, even though we cannot see any means by which He can accomplish His promises to us.

A comet attracts more attention than the steady star, but the star shines on after the comet has passed.

Stars teach us lessons of steadfastness.

Stars differ in glory. We may not all occupy the same place in God's work, either here or in the future life. But each star has its place, and it takes millions of faint stars to make the glory of the Milky Way.

Planets shine by reflected light. "Every gleam of thought, every flash of the intellect, is from the Light of the world."—*Ellen G. White*.

STONES,—See also Rocks.

A fool's wrath is heavier than a stone. Prov. 27:3.

The heart of the wicked is like stony ground. Eze. 11:19; Matt. 13:1-6.

Our daughters are to be like cornerstones. Ps. 144:12.

Stones are symbols of righteousness. Zech. 9:16; 1 Pet. 2:4, 5.

Peter was a rolling stone,—unstable.

STORMS

To make the soul great, God gives us great storms, great dangers to meet, great obstacles to conquer, great deserts to cross.

Deserts, darkness, pestilences, cold, hail, and tempest, all are His agents to purify, strengthen, and enoble the character.

Wind and rain purify the air, without them the air would become stagnant and full of smoke and dust and foul gases.

> "I do not fear for thee, though wroth
>   The tempest rages through the sky;
> For are we not God's children both,
>   Thou, little sandpiper, and I."
>   —*Celia Thaxter.*

Sometimes agitation of disputed questions clears the atmosphere of suspicion and doubt, and enables us to see the mountain peaks of truth more clearly.

STREAMS,—See also Brooks; Rivers.

Streams are used in the Bible as symbols of—
1. Righteousness. Amos 5:24.
2. Glory of the Gentiles. Isa. 66:12.
3. Grief. Ps. 124:4.
4. Deceit. Job 6:15.

STUBBLE

Stubble is a symbol of the wicked. Obadiah 18; Nahum 1:10; Mal. 4:1.

It also represents worthless character. 1 Cor. 3:10-15.

SUMMER

If summer glory reigns in the sky, nothing in the earth can be hidden from its heat. The love of God must eventually fill the earth, and nothing can restrain it from its ultimate purpose.

Summer showers are like tears of gladness. The atmosphere is so full of the richness of the summer-time that it sheds its goodness in pattering rain. And the earth shares in the relief of the skies, as it gently sends up its delicate wisps of vapor after the storm.

SUN

The sun is used as symbolic of,—

1. Christ. Mal. 4:2; Ps. 84:11.
2. The church. Cant. 6:10.
3. The glory of Christ and the angels. Matt. 17:1, 2; 28:1-3; Rev. 1:12-16.
4. The saints. Judges 5:31; Cant. 6:10; Matt. 13:43.

Though the sun is the source of light, we gain nothing but blindness by gazing directly at it. The light of the sun illuminates objects about us, and by using its light we become acquainted with things. Even so we cannot see God himself, but may gain knowledge from a study of His works which reflect His glory.

"The sun bears itself without partiality in infinite abundance and continuity. It is a life-giving stimulus to all things. And it is the emblem of God, of whom it is said; 'He maketh His sun to rise on the evil and on the good, and sendeth rain on the just and on the unjust.'"—*Beecher.*

SUNBEAMS

The Thracians represented God by a sun with three beams—one melting a sea of ice, another melting a rock, and a third putting life into a dead man.

As the flower bulb is drawn upward by the influence of the sunbeams, so our hearts, though buried in worldliness and selfishness, are drawn out into beauty by the power of Christ.

If sunbeams are powerful enough to give the earth a new beauty, what can God's love do in the soul that receives Him?

SUNDEW,—See Plants.

SUNDIAL

Let a man at midnight examine a sundial, and even the brightest light will make it tell only falsehood. The Word of God can be understood only by the illumination of the Holy Spirit. Human intellect, no matter how brilliant, will give nothing but erroneous conclusion.

SUNLIGHT

As the sunlight produces its effect on nature, so thoughts and impressions are stamped on minds of children.

Landscapes are made by a mingling of light and shadow. In human life there is mingled joy and sorrow. As the shadows make the highlights more prominent, so sorrows make joys the more worth having.

It is the continuous sunlight day after day that brings the harvest. Make it as bright as you please, you could not get heat and light enough in one day to do any good. A Christian must dwell in the light, not merely expose himself to a casual influence of God's grace.

It is possible to filter the heat from a sunbeam and allow only the cold light to pass through. So one may analyze and criticize the works of God in such a way as to cut off the warmth of His love, and let only the cold light of scientific data pass into the mind.

A beam of sunlight in a room will reveal the smallest particle of dust floating in the air. So a ray of divine light reveals the smallest imperfections in our lives.

The morning light comes gradually and quietly over the earth. So God's mercy comes into the soul.

SUNRISE

> "I'll tell you how the sun rose,—
>   A ribbon at a time.
> The steeples swarm in amethyst,
>   The news like squirrels ran.
>
> The hills untied their bonnets,
>   The bobolinks begun
> Then I said softly to myself,
>   'That must have been the sun!'"
>   —*Emily Dickinson.*

If God makes the day so beautiful before He gives it to us, how sacredly we should guard it, lest any wrong thought or word or deed of ours should mar it.

SUNSET

Often when walking in the woods at sunset, we cannot see the sun, but we know that it is shining, from the brightness of the leaves above our heads. So with the Holy Spirit; we cannot see Him, but we know He is near, for His work is manifest in the lives of Christians.

"Lamps so heavenly must have been lighted from on high."

Among the Alps, when darkness creeps over the valleys, Mont Blanc rises above the shadows, catching the retreating sun, and is flushed with a rose-colored glow that is beyond words to describe. So in life, even though the darkness of selfishness or anger hides the valleys of the soul, the beauty of past favors and kindnesses glows with a radiance that seems to forbid the advance of night any farther.

> "God walks among the darkening hills,
> When the sun in the west sinks low,
> For I have felt His presence there,
> And that is how I know."
> —*Nathanael Krum.*

The setting sun is like a man whose influence remains after he has gone.

SWALLOW,—See Birds.

SWIFTS,—See Birds.

TARES,—See also Weeds.

Tares represent wicked men. Matt. 13:38-40.

Like tares the wicked will be destroyed, even though they grow together with the wheat until the harvest.

TEETH

Teeth of horses continue to grow until late in life. As the coarse herbage wears off the surface, new growth is pushed up from the roots.

Gnawing animals, such as ground squirrels, rats, mice, and squirrels have two sharp cutting teeth in front on each jaw. These are made of soft enamel behind a layer of hard enamel. As the soft enamel wears away, it leaves a sharp cutting edge of hard enamel in the front.

THORNS

The wicked are compared to thorns. Num. 33:55; 2 Sam. 23:6; Micah 7:4.

A thorn is a degenerate structure, the result of unfavorable conditions. Yet it is not entirely useless, as it protects the plant. God has so made the world that even the results of sin shall serve a useful purpose whenever possible.

In the affairs of men and nations, He makes the wrath of man to serve Him. He is able to foresee all their plans and to turn their evil devices to accomplish His will.

Paul spoke of his affliction as a "thorn in the flesh." It kept him from becoming proud. So we may sometimes have petty irritating circumstances; but if we learn to endure them patiently, we may be the better for them.

Usually the thorn in the flesh causes us to stop and remove the offending object. There is no need to suffer needlessly, for sheer lack of initiative, that which might easily be remedied.

TIN

Tin is used as a symbol of sin. Isa. 1:25; Eze. 22:18-20.

It might also be used to symbolize the protection that God gives His children. Tin plating covers the iron, and prevents its rusting.

TIDE

When the tide is low, every little pool is a sea by itself. The tiny shrimp in one pool have nothing in common with those in another. Like tiny minds and souls they fancy that their pool is the only one there is. But when the tide comes in, all the pools are lost in the greatness of the sea. When the tide of rising thought and broadening intellectual life sweeps away the restrictions of human intolerance, then all may come together in the greater fellowship of mind and spirit.

A landlocked bay will be calm and peaceful when the tide is full, but when it is low, and the currents sweep through the channels, there are muddy whirls and hidden sandbars and dangerous eddies. When the heart is filled with the Holy Spirit, all is calm and quiet, but when the spiritual life is at low ebb, there are dangerous currents and eddies in the life. *(Henry Ward Beecher.)*

TOAD,—See Animals.

TREES,—See also Flowers; Fruit; Leaves; Plants; Saplings.

Wisdom is a tree of life. Prov. 3:13-18.

The life of the righteous is like the days of a tree. Isa. 65:22.

Trees represent the righteous. Ps. 1:1-3; 92:12-14; Isa. 61:3; Jer. 17:8.

Deformed trees are like youth who have developed bad habits. They are likely to retain the deformity throughout life.

Trees may be grafted. We need to be grafted on to Christ in order to bear fruit.

When a scion from one tree is grafted on to the stock of another the two cambium layers must be brought into contact, in order that the life may go from one to another. When we are grafted into Christ, we must maintain an active connection with His life, in order that we may grow.

"The words of men are like leaves of the tree; when there are too many, they hinder the growth of the fruit."

Some trees are anchored by tap roots that go deep into the earth. In time of storm, when other trees go down, they stand erect. If we would enjoy in later life a stability of character that will stand the fierce gusts of temptation, we must strike our roots deep into the truth early in life. Like a tree, we cannot develop a strong tap root in haste, while the tempest is forcing us down.

"The groves were God's first temples."—*Bryant*.

The largest Sequoia originates in a single cell so small that it requires a powerful microscope to make it visible. The power of God continuously exerted, develops this tiny cell into the mighty giant of the forest.

Trees begin preparation to shed their leaves long before cold weather comes. The seasonal rhythm has become established, and they respond to it. Maple, beech, apple, and pear trees transported to warmer lands continue to drop their leaves, even though it does not turn cold. The habit has become so ingrained in their nature that they cannot change.

"What a thought that was, when God thought
of a tree!"

In the Golden Gate Park in San Francisco are many flowering trees and shrubs imported from South Africa.

These bloom in September and October, which are the spring months in their native land. They are following the laws of their life, the rhythm that has become established in them.

> "Teach me, Father, how to be
> Kind and patient as a tree."
> —*Edwin Markham.*

A tree is as perfect in one stage of its existence as at another. It is a perfect seed, a perfect seedling, a perfect sapling; its youth and its old age are perfect. Perfection consists in following the plan of the Creator in each stage of its development. "Be ye therefore perfect,"—perfect children, youth, and mature men and women.

Some trees wither at the first blast of cold air. They are like sensitive people that can not endure adversity or hardship. But the more sturdy ones live on, untouched by storms.

> "But only God can make a tree."—*Kilmer.*

Lessons from specific trees:

Birch,—Probably no more lovely sight is to be found in nature than a grove of white birches. The graceful trunk and branches, the shining bark, the delicate leaves, all blend into a picture of grace and delicate loveliness. Birches usually grow best in the cool northern forests, where they help compensate for the lack of some other ornamental trees found farther south.

Buckeye,—The buckeye produces a long spike of blossoms, but only one or two at the end of the spike ever form fruit. The rest are simply for beauty. These symbolize the way God loves to do pleasant things.

Cedar,—The Bible uses cedar to represent,—
1. Christian growth. Ps. 92:12-14.
2. Great men. Eze. 31:3-18.

Chestnut,—Uncourteous, rough people prick like a chestnut bur. Yet many of them contain good nuts. Even a rough exterior may hide a good heart.

Elm,—Some trees lose their beauty when the leaves are gone, but the elm tree stands boldly forth in perfect symmetry against the wintry sky. Some men are like the elm; their character shows out best when the storms of adversity have stripped them of the externals of life, and left their inner soul exposed.

Evergreens,—Evergreen trees are like Christians who are not moved by persecutions and evil influences. Ps. 104:16.

Maple,—The sugar maple yields a rich harvest of sweetness each spring to the farmer who will take its sap and boil it down. But no one could obtain the sugar by cutting the tree and boiling its wood, the sap must be collected, drop by drop as it flows. Many of God's blessings are this way. We must receive them drop by drop as they flow from His storehouse. If in our greed we try to seize everything at once, we meet with failure.

The sugar maple will give its sweet sap only when the nights are cold enough to freeze. It is the alternate freezing and thawing that keeps up the flow. When the weather becomes so warm that the buds start, the sap is of no further use in sugar-making. Sometimes we wonder why we have to undergo alternate freezing and thawing in our experience; but perhaps God is trying by that method to extract the sweetness that we might otherwise never give.

Oak,—The oak is a symbol of strength. Amos 2:9.
Dying oaks are like sinners. Isa. 1:26-30.
"Acorns and graces sprout quickly, but grow a long time before ripening."—*Henry Ward Beecher.*
The fiercer the storm, the deeper the oak strikes its roots into the earth.

Olive,—Aged and decayed olive trees are surrounded by young shoots growing from their roots. Thus do children gather about aged and infirm parents. Ps. 128:3.

Palm,—Palms are used as a symbol of the righteous. Ps. 92:12.

When the sago palm is young, it is covered with sharp thorns to protect it from enemies, as it grows older, the thorns drop off. So does a young Christian need to possess an impenetrable barrier against those who would challenge his faith. But as faith and love grow, and experience is gained, the firmness and strength of maturity become a safeguard, and less severe measures need be taken to preserve the integrity.

As the palm brings forth beauty and fruit in the desert, so God's children are to bear the fruits of righteousness in this world.

Pine,—The pine tree is the most widely distributed of any tree. It is a symbol of the north temperate zone, and represents strength and patient endurance in the face of difficulty.

It grows on bare steep mountain slopes, where gray skies and fierce gales surround it. Yet in spite of wind, hunger, and cold, it becomes a thing of beauty.

Not only is the pine tree beautiful, but it serves a practical purpose. It condenses the fog and distils its moisture to the earth below. Its needles filter the moisture slowly through, and releases it in steady streams to the valleys.

"The mightiest rivers are cradled in the leaves of the pine."—*Chinese Proverb*.

Cones are all built of spiral rows of scales. The basic pattern is extremely simple. Yet by varying the length of the cone, the size of the scales, and adding a few decorative features, God has made such a simple device as a cone a feature of great beauty.

God has fitted the pine for its life by giving it shallow roots to grasp the rocks, thick bark to resist the cold, and needles that live through the winter in spite of the cold.

Redbud,—In this tree the brilliant crimson blossoms appear before the leaves. The flaming beauty of the flowers attracts many insects. But every bee that alights to gather honey imbibes a fatal poison, and drops dead. So the flaming crimson flowers of this earth are fatal to those who would enjoy their sweetness.

Willow,—The alpine willow forms a mat upon the ground in high altitudes where no other tree or shrub can grow. Its catkins project straight up about two inches above the surface to the matted stems.

OF NATURE

TURTLES,—See Reptiles.

UNITY

Unity does not mean uniformity. There is none; there can be none in the free universe of God. There is no uniformity in nature. Wherever life exists there will be found a variety of forms and activities. Unity consists in harmonious and sympathetic cooperation between all these diverse elements.

VAPOR,—See also Dew; Mist; Rain; Snow.

Man's life is a vapor. James 4:14.

Water vapor rising from the ocean, lakes, and rivers is carried over the land in gentle, harmless form. There it is condensed and falls as drops of rain or soft flakes of snow. How much better this is than to bring it up over the land in torrents or waves that might wreck and destroy.

VENUS FLY-TRAP,—See Plants.

VINE,—See Plants.

VIOLET,—See Plants.

VOICE

"And God said" "By the word of the Lord were the heavens made, and all the host of them by the breath of His mouth." Ps. 33:6.

When God spoke first in this world His word turned into flowers. His speech was made visible to us. He is still speaking to us in His word and His highest revelations reach us in the bloom of the lily that grows in the wooded glen.

The human voice spoken into a radio transmitter will send forth radiations through space that will, when picked up by a suitable instrument, become transformed into exact duplicates of the original tones. Even so God is sending His invisible radiations of power through space. The beauties of nature in their infinite range of color, sound, form, and motion, are but the visible reproductions of His voice. Ps. 29:3-9.

WASP,—See Insects.

WATER,—See also Dew; Mist; Rain; Vapor.

Water is used in many ways in the Scripture. Among its most beautiful symbolisms are the following:

1. Counsel is like deep water. Prov. 20:5.
2. Water represents affliction. Ps. 69:1, 2; 73:10; 124:1-5.
3. The voice of God is like many waters. Ps. 29:3; 93:4; Eze. 43:2.
4. Water represents cleansing from sin. Ps. 51:7-10; Isa. 52:15; Eze. 36:25-29; John 15:3; Heb. 10:19-22.

Living water is the grace of God in the soul.

Water always flows downward, to seek its own level. Of its own power it never would rise. Only the influence of the sunlight to convert it into vapor, can ever raise it to the clouds and bring it in blessing over the land.

When water is cooled, it becomes heavier until it reaches the temperature of 39°. From there to the freezing point, 32°, it expands, and becomes lighter. Thus the cold water floats on the top, and when ice is formed, it floats. If water kept on getting heavier until it froze, ice would form on the bottom of the lakes and streams. Once having formed, it would never melt, for in the summer the warmed surface waters, being lighter, would remain on top, and never reach the bottom unless the whole body of water were warmed through.

WATER LILY.—See Flowers.

WAVES.—See also Billows; Ocean; Sea.

Waves are used in the Bible as a symbol of,—
1. Instability. James 1:6.
2. Sorrow and grief. Ps. 42:7; 88:6, 7.
3. The wicked. Jude 13.
4. Righteousness. Isa. 48:18; 57:20, 21.

Waves only affect the surface. Deep down in the ocean they have no effect. Storm and trouble cannot affect the tranquility of the Christian, for his joy lies beneath the surface.

WEATHER

In dark, gloomy weather the spirits are depressed, but in bright, sunny weather one feels happy and full of optimism. Evil or wrong-doing brings depression, but righteousness brings joy and peace.

WEEDS

A weed is a plant out of place. The best of plants may become a nuisance by persisting in growing where they are not wanted.

Weeds are "successful plants." By their persistence they have established themselves in spite of all attempts to discourage them. Sometimes we need the persistence of weeds to establish ourselves in spite of opposition.

Often the most undesirable appearing weeds have beautiful flowers. Two lessons are readily found in this fact. First, even the most unlikely looking individual may produce something of real beauty. Second, even the most beautiful flowers may hide a miserable and despicable weed.

Some of our most lovely cultivated flowers have come from common weeds. All that was needed was care and attention to bring out their latent powers.

Spirituality is a tender plant, and without great care it soon dies. Sin needs neither hoeing nor watering, but springs up like a weed. *(C. H. Spurgeon.)*

WELLS,—See Springs.

WHALE,—See Animals.

WHEAT,—See also Grain; Seed.

Wheat is a symbol for the death of Christ, for sacrifice, and for the harvest of righteousness. It also represents righteous men. Matt. 3:12; 13:38.

WHIRLPOOL

A person who falls into careless habits is like a boat that is nearing a whirlpool. There may be no indication of danger, but in course of time he recognizes that he is making no progress, only "going in circles." Now is the time to put forth a struggle to escape.

OF NATURE 149

If this is not done, the power of the current increases. Every relaxation allows him to be drawn nearer to the vortex, and at length the dizzy pace sucks him under, a shipwreck of conscience, of reputation and of everything of value in human character.

WHIRLWIND
God's wrath is like a whirlwind.
Strife among nations is a whirlwind that sweeps everything into its raging vortex of destruction.

WILLOW,—See Trees.

WIND,—See also Air; Weather.
Words of the desperate are as wind. Job 6:26, 27; 8:1, 2.
The life is as wind. Ps. 78:38-40.
Wicked are as stubble before the wind. Job 21:16-18; Ps. 1:4.
Terrors are as wind. Job 30:15.
A contentious woman is like the wind. Prov. 27:15, 16.
Wind is a symbol of the Holy Spirit. John 3:8; 20:21, 22; Acts 2:1, 2.
It represents trial and temptation. Matt. 7:24-27.
Strife and commotion is like a wind. Rev. 7:1-3; Dan. 7:1-3.

Wind is moving air. Its movement carries away impurities and brings fresh supplies of pure air.

Wind gives grace and beauty to the landscape by moving the grass, flowers, and trees. Nothing is more impressive than a field of waving grain.

Wind sweeps the earth of fallen leaves, piling them up

where they are wet by the rain and reduced to soil elements by the action of bacteria.

"The wind bloweth where it listeth." Yet it is not erratic, but is under the rule of law. God's laws are not capricious, even though we cannot always understand them.

Satan is the "prince of the power of the air." By God's permission he may bring storms to destroy.

> "The winds my tax-collectors are,
>   They bring me tithes divine,—
> Wild scents and subtle essences,
>   A tribute rare and free."
> —*Lucy Larcom.*

WINE,—See also Bread.

Unfermented wine is a symbol of the purity of Christ. It represents His blood shed for us.

WINTER

A certain amount of cold and snow are essential to the health and vigor of some plants. If it were always summer, they would die. We would miss many of the finer traits of Christian character if it were not for the influence of adversity and trouble.

WOLF,—See Animals.

WOODS

> "So I stole from the crowd to the heart of the wood;
>   I trod where the Master trod;
> I entered His tabernacle there,
>   And communed a while with God."
> —*Nathanael Krum.*

WOODPECKER,—See Birds.

WOOD-SORREL

When the day is cloudy or dull, the foliage of the oxalis droops and closes. In the sunshine it displays its delicate beauty of leaf and blossom.

Sensitive minds close up under the influence of rough, uncouth treatment, but open when warmed by love and kindness.

WORDS,—See also Orange; Voice; Speech.

When we are attacked by scandalous or lying words, it is better to yield than to answer them in kind. "A soft answer turneth away wrath."

WORMS

Man is represented as a worm. Job 25:6; Ps. 22:6; Isa. 41:13, 14.

YEAST,—See Leaven.

THE END

www.ingramcontent.com/pod-product-compliance
Lightning Source LLC
Chambersburg PA
CBHW071207160426
43196CB00011B/2220